AF388902

Biosensors with
Magnetic Nanocomponents

Biosensors with Magnetic Nanocomponents

Editor

Galina V. Kurlyandskaya

MDPI • Basel • Beijing • Wuhan • Barcelona • Belgrade • Manchester • Tokyo • Cluj • Tianjin

Editor
Galina V. Kurlyandskaya
Universidad del Pais Vasco—Euskal Herriko Unibertsitatea
Spain

Editorial Office
MDPI
St. Alban-Anlage 66
4052 Basel, Switzerland

This is a reprint of articles from the Special Issue published online in the open access journal *Sensors* (ISSN 1424-8220) (available at: https://www.mdpi.com/journal/sensors/special_issues/Magneticnano).

For citation purposes, cite each article independently as indicated on the article page online and as indicated below:

LastName, A.A.; LastName, B.B.; LastName, C.C. Article Title. *Journal Name* **Year**, *Article Number*, Page Range.

ISBN 978-3-03936-680-4 (Hbk)
ISBN 978-3-03936-681-1 (PDF)

Cover image courtesy of Galina Kurlyandskaya.

Contents

About the Editor . . . . **vii**

Preface to "Biosensors with Magnetic Nanocomponents" . . . . **ix**

Mohammad Reza Zamani Kouhpanji and Bethanie J. H. Stadler
A Guideline for Effectively Synthesizing and Characterizing Magnetic Nanoparticles for Advancing Nanobiotechnology: A Review
Reprinted from: *Sensors* **2020**, *20*, 2554, doi:10.3390/s20092554 . . . . **1**

Nikita A. Buznikov and Galina V. Kurlyandskaya
Magnetoimpedance in Symmetric and Non-Symmetric Nanostructured Multilayers: A Theoretical Study
Reprinted from: *Sensors* **2019**, *19*, 1761, doi:10.3390/s19081761 . . . . **39**

Zhen Yang, Anna A. Chlenova, Elizaveta V. Golubeva, Stanislav O. Volchkov, Pengfei Guo, Sergei V. Shcherbinin and Galina V. Kurlyandskaya
Magnetoimpedance Effect in the Ribbon-Based Patterned Soft Ferromagnetic Meander-Shaped Elements for Sensor Application
Reprinted from: *Sensors* **2019**, *19*, 2468, doi:10.3390/s19112468 . . . . **53**

Yapeng Zhang, Jingjing Cheng and Wenzhong Liu
Characterization and Relaxation Properties of a Series of Monodispersed Magnetic Nanoparticles
Reprinted from: *Sensors* **2019**, *19*, 3396, doi:10.3390/s19153396 . . . . **65**

Felix A. Blyakhman, Sergey Yu Sokolov, Alexander P. Safronov, Olga A. Dinislamova, Tatyana F. Shklyar, Andrey Yu Zubarev and Galina V. Kurlyandskaya
Ferrogels Ultrasonography for Biomedical Applications
Reprinted from: *Sensors* **2019**, *19*, 3959, doi:10.3390/s19183959 . . . . **81**

Aleksandr Ryzhkov and Yuriy Raikher
Size-Dependent Properties of Magnetosensitive Polymersomes: Computer Modelling
Reprinted from: *Sensors* **2019**, *19*, 5266, doi:10.3390/s19235266 . . . . **95**

Marcus Vinicius Lopes, Edycleyson Carlos de Souza, João Gustavo Santos, João Medeiros de Araujo, Lessandro Lima, Alexandre Barbosa de Oliveira, Felipe Bohn and Marcio Assolin Correa
Modulating the Spin Seebeck Effect in Co$_2$FeAl Heusler Alloy for Sensor Applications
Reprinted from: *Sensors* **2020**, *20*, 1387, doi:10.3390/s20051387 . . . . **107**

Alfredo García-Arribas
The Performance of the Magneto-Impedance Effect for the Detection of Superparamagnetic Particles
Reprinted from: *Sensors* **2020**, *20*, 1961, doi:10.3390/s20071961 . . . . **121**

Gabriele Barrera, Marco Coisson, Federica Celegato, Luca Martino, Priyanka Tiwari, Roshni Verma, Shashank N. Kane, Frédéric Mazaleyrat and Paola Tiberto
Specific Loss Power of Co/Li/Zn-Mixed Ferrite Powders for Magnetic Hyperthermia
Reprinted from: *Sensors* **2020**, *20*, 2151, doi:10.3390/s20072151 . . . . **131**

Beatriz Sisniega, Ariane Sagasti Sedano, Jon Gutiérrez and Alfredo García-Arribas
Real Time Monitoring of Calcium Oxalate Precipitation Reaction by Using Corrosion Resistant
Magnetoelastic Resonance Sensors
Reprinted from: *Sensors* **2020**, *20*, 2802, doi:10.3390/s20102802 . . . . **147**

About the Editor

Galina V. Kurlyandskaya, IEEE Senior Member, graduated from the Physics Department of Ural State University A.M. Gorky, Ekaterinburg, Russia, in 1983. She started her research work in 1983 at the Institute of Metal Physics UD RAS. She obtained her PhD in physics of magnetic phenomena in 1990 and her Doctor of Science degree in 2007 from Ural State University A.M. Gorky. Prof. Kurlyandskaya received advanced training at the Institute of Applied Magnetism, University of Complutense, University of Oviedo, University of the Basque Country, Euskal Herriko Unibertsitatea UPV/EHU, University of Dusseldorf Heinrich Heine, ENS Cashan, University of Maryland, Rowan University, Immanuel Kant Baltic Federal University, Ural State University A.M. Gorky, Ural Federal University B.N. Yeltsin (Laboratory of Magnetic Sensors), Brazilian Research Center of Physics (CBPF) and University of Santa Maria. Her main research areas are fabrication and the magnetic and transport properties of nanostructured magnetic materials, magnetic domain structure, magnetoabsorption, magnetic sensors and biosensors, and biomedical applications of magnetic nanocomposites.

Prefaceto "Biosensors with Magnetic Nanocomponents"

The book you have in your hands is a result of the special efforts of an international team from Brazil, China, Italy, Russia, Spain, and the United States of America. The works of all the members of this multidisciplinary team are especially appreciated, as the last 5 months of the Issue coincided with the coronavirus world tragedy. We all learned from this new experience and started to realize the need for extra efforts in the field of biomedical applications and public health. This book contains peer-reviewed contributions from the Special Issue "Magnetic Materials Based Biosensors" in MDPI's Sensors. The works herein were submitted to the journal in the period from February 2019 to June 2020. This book contains nine research works and one topical review. PhD students, researchers, and the educational community working in the fields of magnetic nanomaterials and biomedical applications of nanocomposites with magnetic components will find this book useful.

The selective and quantitative detection of biocomponents is greatly requested in biomedical applications, clinical diagnostics, and the development of new composites from biotechnological roots. On one hand, many traditional magnetic materials are not suitable for the ever-increasing demands of these processes. On the other hand, the list of requested applications is rapidly growing. The push for a new generation of microscale sensors for biomedical applications continues to challenge the materials science and engineering communities to work together in close collaboration with medical teams aiming to develop novel compact analytical devices that are suitable for such purposes.

The principal requirements of a new generation of nanomaterials for sensor applications are based on well-known demands: high sensitivity, small size, low power consumption, stability, quick response, resistance to aggressive media, low price, and easy operation by nonskilled personnel. In addition, the possibility of integration of on-chip sensitive elements with nanoscale components is also expected for the next generation of devices.

There are different types of magnetic effects capable of creating sensors for biology, medicine, and drug delivery, including magnetoresistance, spin valves, Hall and inductive effects, magnetoelastic resonance, and giant magnetoimpedance. Although many geometries are still under testing, thin films and nanostructured multilayers are preferable, as they are most compatible with semiconductor electronics and existing electronic circuit fabrication technologies. There are different reasons contributing to the delay of the competitive integration of high-frequency nanostructured thin film elements into the global market. One of them is the need for additional understanding of basic concepts of microwave radiation absorption by nanostructured multilayered elements. Another is the need for the elaboration of simple, fast, and cheap characterization of materials with high dynamic permeability.

The present goal is to design nanomaterials both for magnetic markers and sensitive elements as synergetic pairs working in one device with adjusted characteristics of both materials. Synthetic approaches using the advantages of simulation methods and synthetic materials mimicking natural tissue properties can be useful, as can the further development of modeling strategies for magnetic nanostructures.

In fact, one of the most interesting cases greatly requested for cancer therapies, the detection of magnetic nanoparticles incorporated into biological tissues, has not been yet properly addressed. Biological tissues present a huge variety of morphologies, and therefore the development of magnetic biosensors is conditioned by the fabrication of reliable samples. One of the strategies for solving this problem is to substitute biological samples at a certain stage of the development of the biosensor by synthetic hydrogel (experimental model of the cytoskeleton) with a certain amount of magnetic nanoparticles, which is capable of mimicking the main properties of living tissues. Here, special attention was also paid to the remarkable multimodal properties of magnetic nanoparticles, as they are very important in resolving challenges slowing the progression of biotechnology.

Galina V. Kurlyandskaya
Editor

Review

A Guideline for Effectively Synthesizing and Characterizing Magnetic Nanoparticles for Advancing Nanobiotechnology: A Review

Mohammad Reza Zamani Kouhpanji [1,2] and Bethanie J. H. Stadler [1,3,*]

1   Department of Electrical and Computer Engineering, University of Minnesota, Minneapolis, MN 55455, USA; zaman022@umn.edu
2   Department of Biomedical Engineering, University of Minnesota, Minneapolis, MN 55455, USA
3   Department of Chemical Engineering and Materials Science, University of Minnesota, Minneapolis, MN 55455, USA
*   Correspondence: stadler@umn.edu

Received: 9 April 2020; Accepted: 26 April 2020; Published: 30 April 2020

Abstract: The remarkable multimodal functionalities of magnetic nanoparticles, conferred by their size and morphology, are very important in resolving challenges slowing the progression of nanobiotechnology. The rapid and revolutionary expansion of magnetic nanoparticles in nanobiotechnology, especially in nanomedicine and therapeutics, demands an overview of the current state of the art for synthesizing and characterizing magnetic nanoparticles. In this review, we explain the synthesis routes for tailoring the size, morphology, composition, and magnetic properties of the magnetic nanoparticles. The pros and cons of the most popularly used characterization techniques for determining the aforementioned parameters, with particular focus on nanomedicine and biosensing applications, are discussed. Moreover, we provide numerous biomedical applications and highlight their challenges and requirements that must be met using the magnetic nanoparticles to achieve the most effective outcomes. Finally, we conclude this review by providing an insight towards resolving the persisting challenges and the future directions. This review should be an excellent source of information for beginners in this field who are looking for a groundbreaking start but they have been overwhelmed by the volume of literature.

Keywords: magnetic nanoparticles; nanobiotechnology; nanomedicine; therapeutics; biosensing

---

1. Introduction

Advancement of nanotechnology has extensively expedited the emergence of novel magnetic nanostructures by reducing the dimensions to 2D nanomaterials, such as thin films and supperlattices, or 1D nanomaterials, such as magnetic nanowires (MNWs), and even 0D, such as spherical magnetic nanoparticles. The excellent quantum efficiency achieved using these nanomaterials has made them useful building blocks for diverse research areas, including medical treatment [1–5], environmental science [6,7], and quantum devices [8–11]. These magnetic nanostructures have opened numerous opportunities for scientists in different disciplines such as nanomedicine, molecular biology [12–14], applied physics, and nanostructured materials [15–20].

Among all magnetic nanostructures, the low dimension magnetic nanostructures, 0D and 1D magnetic nanoparticles, have attracted huge attention over the last few decades as they provide multimodal functionality priming multitude aspects of the nanomedicine and therapeutics applications. As the magnetic nanoparticles' dimensions and size are reduced, due to the competition between the magnetic energies, in addition to their composition, the magnetic nanoparticles present different magnetic behaviors, such as ferromagnetic, superparamagnetic, and ferrimagnetic (see Figure 1). The ferromagnetic and superparamagnetic nanoparticles are opposite, as the former ones have long range ordered magnetic moment leading to have non-zero magnetization at zero fields, while the latter do not possess a stable magnetic moment, due to thermal fluctuations, leading in zero magnetization at zero fields. Note that ferrimagnetic nanoparticles are an intermediate state between these two states where they would be superparamagnetic if their sizes are sufficiently small so that no domain walls can be formed.

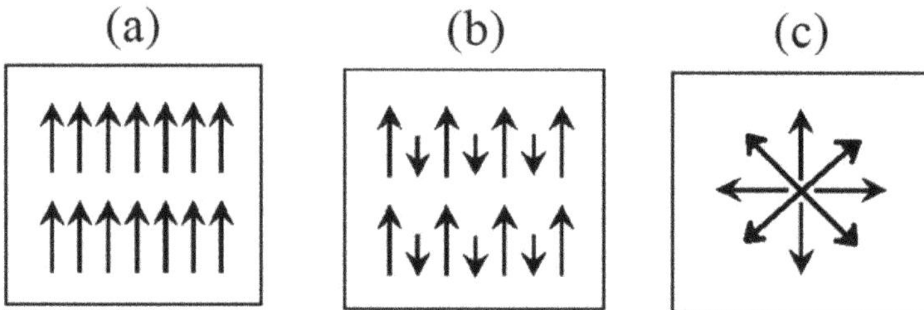

Figure 1. Schematic illustrating the dominant magnetic moment configurations in nanoparticles: (a) ferromagnetic: summing up long range ordering of magnetic moments, (b) ferrimagnetic: subtracting long range ordering of magnetic moments, opposite directions in the neighboring domains, and (c) superparamagnetic: continuous fluctuations of the magnetic moment leading to a net zero magnetic moment.

The multimodal functionality of the magnetic nanostructures requires an accurate and precise characterization of these nanostructures, which may inhibit or enhance their use depending on the application. Unfortunately, the high yielding nanofabrication processes for magnetic nanoparticles do not allow perfectly identical production, leading to variation in their magnetic characteristics and functionalities, ultimately inefficient for the proposed application. Consequently, in order to suppress this persistent challenge, it is important to understand the strengths and weaknesses of the synthesis processes as they are fundamental for producing identical magnetic nanoparticles with unique properties. Furthermore, understanding the reliability and validity ranges of the diverse characterization techniques is crucial for determining the most effective synthesis process for achieving magnetic nanoparticles with desired properties for a particular bio-application. In this review, we briefly provide details regarding the most commonly used synthesis processes to realize the most effective approach for tailoring magnetic nanoparticles. We then explain the diverse techniques used for characterizing the size, morphology, composition, and magnetic properties of these magnetic nanoparticles. Finally, we provide objective recommendations for selecting the most effective synthesis approach for producing magnetic nanoparticles for specific applications.

2. Synthesis Processes

Magnetic nanoparticles can be divided into two groups based on their dimensions: 0D and 1D. Each category can be further divided into sub-categories based on their shapes or aspect ratios, defined as the ratio of the longitudinal size to the lateral size. For example, 1D magnetic nanoparticles include nanodiscs, which are particles with aspect ratios equal or smaller than one and nanowires with aspect ratios larger than five. The magnetic properties of the magnetic nanoparticles determine the best synthesis path. For example, the 1D ferromagnetic magnetic nanoparticles are mainly fabricated using either the template-assisted method or template-free methods. In both categories, the flux of ions can be produced using several methods, such as chemical vapor deposition, physical vapor deposition, atomic layer deposition, laser pulse deposition, and electrochemical deposition. Except for electrochemical deposition, other techniques are not very common in the fabrication of the magnetic nanowires because they require high energy and vacuum pressure that are costly. Furthermore, in addition to the very low yields of these techniques, they also suffer from uniform growth of the magnetic nanowire, especially if high aspect-ratio magnetic nanoparticles over 1000, such as in template-assisted electrodeposition of magnetic nanowires, are desired. Therefore, here we focus on the electrochemical deposition technique that requires a template for synthesis. To date, numerous methods for synthesizing the magnetic nanoparticles have been proposed and successfully employed for the fabrication of diverse magnetic nanoparticles. Considering the cost and controllability of size/shape, all these synthetic methods can be categorized into two main categories according to the used solvent: aqueous or non-aqueous solvents [21]. The aqueous-based magnetic nanoparticles are relatively cheap; however, controlling their sizes and shapes is very challenging. The non-aqueous-based methods provide good control of the size and shape while they are relatively more expensive compared to the aqueous-based methods. Here, we provide a brief review of the most popular synthetic methods.

2.1. Co-Precipitation

Co-precipitation is the most commonly used approach that can be done at room temperature or elevated temperature (Figure 2). The solution consists of mixing ferrous and ferric ions in a molar ratio of 2:1 protected using an inert gas. In this method, the solution pH is a very important factor as a lower pH is desirable for nucleation of the Fe_3O_4 nucleus while the higher pH facilitates the growth of the Fe_3O_4 nucleus. The capability of this method for mass-production of magnetic nanoparticles has placed in a central position leading to several attempts to modify this method to enhance the magnetic nanoparticles' magnetic properties and morphology [22]. For example, Wu et al. employed ultrasonic-assisted chemical co-precipitation to achieve magnetic nanoparticles with a nominal size of 15 nm with high purity [23]. Another example is the work by Pereira et al. where they synthesized magnetic nanoparticles with a nominal size of 5 nm using a one-step aqueous co-precipitation that employs alkanolamines [24]. These two examples represent a significant technological development as they are capable of mass producing magnetic nanoparticles with reduced average size while enhancing the magnetization moment. The size and shape control of the magnetic nanoparticles using this technique is very challenging, and furthermore, the presence of multi-phase magnetic nanoparticles is a common limitation [25]. The mass production of magnetic nanoparticles with large magnetization saturation usually suffers from particle aggregation. To overcome this limitation, a coating is essential, which was shown to readily be done using Ag and Au or introducing ligands.

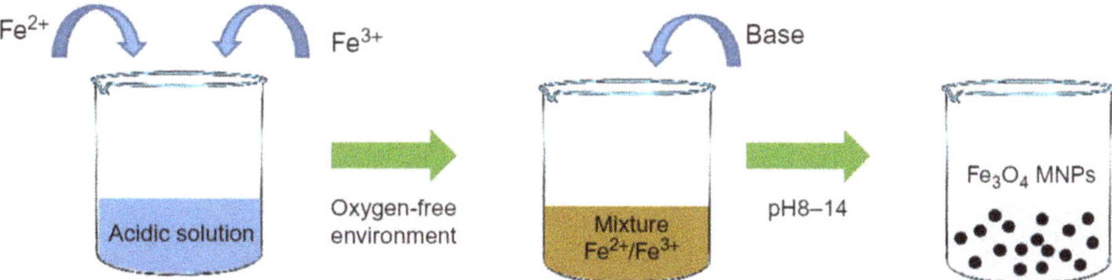

Figure 2. A schematic of the synthesis of iron oxide magnetic nanoparticles using the co-precipitation method [26]. In this example, the precursors (Fe^{2+}/Fe^{3+} chlorides, sulfates, or nitrates) are dissolved in an acidic solution. Then, a strong base is added to increase the pH > 8 in a non-oxidizing environment.

2.2. High-Temperature Thermal Decomposition

The thermal decomposition approach overcomes the size and morphology disparities limitation of the co-precipitation method. Generally speaking, the magnetic nanoparticles synthesized at higher temperatures provide more uniform size distributions [27]. The high-temperature decomposition also provides a route towards more crystalline magnetic nanoparticles (Figure 3). The main advantage of this method over co-precipitation is that it decouples the nucleation and growth of the magnetic nanoparticles leading to monodisperse, narrow size distribution, and highly crystalline magnetic nanoparticles [21]. It is possible to incorporate the inexpensive and non-toxic iron chloride to produce monodisperse magnetic nanoparticles without the need for size selection processes [28]. This approach also has the ability to control the crystallinity in ways suitable for producing various shapes, such as nano-cubic and nano-octahedral shapes [29]. Nevertheless, the magnetic nanoparticles synthesized using the thermal decomposition technique, especially those synthesized on aqueous media, tend to degrade in long term which makes their clinical applications debatable.

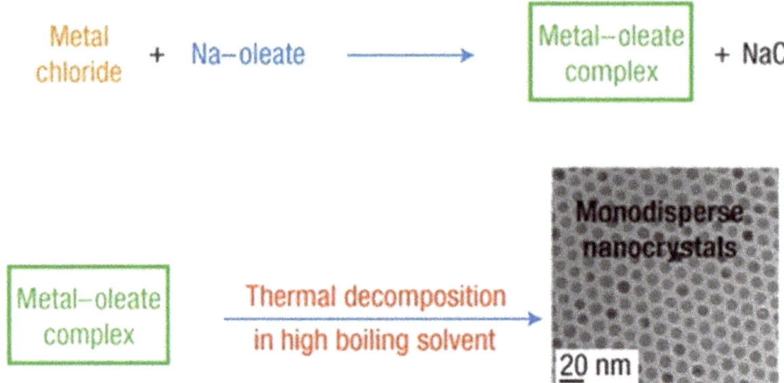

Figure 3. A schematic for magnetic nanoparticles preparation using thermal decomposition technique. Metal–oleate precursors were prepared from the reaction of metal chlorides and sodium oleate. The thermal decomposition of the metal–oleate precursors in the high boiling solvent produced monodisperse nanocrystals [30].

2.3. Hydrothermal and Solvothermal Synthesis

Hydrothermal and solvothermal syntheses employ various wet-chemical techniques to form crystalline magnetic nanoparticles. Figure 4 shows a schematic of the hydrothermal method, where the process is carried out in high-pressure reactors or autoclaves to reach high pressures at high temperatures. This method uses either aqueous or non-aqueous solutions at high temperatures under high pressures to avoid the growth of dislocations in single crystal magnetic nanoparticles [30]. As a result, this method is suitable for the growth of crystalline phases that are unstable around their melting temperature. Furthermore, this method facilitates the growth of the magnetic nanoparticles that have a very high vapor pressure at their melting points while maintaining good control over the magnetic nanoparticles' compositions [31]. This method is especially beneficial for the synthesis of hollow and controlled shape magnetic nanoparticles [32] including nanotubes and nanorings. It should be mentioned that this technique is very sensitive to the synthesis temperature as it can dramatically impact the reaction kinetics and nucleation rate [33].

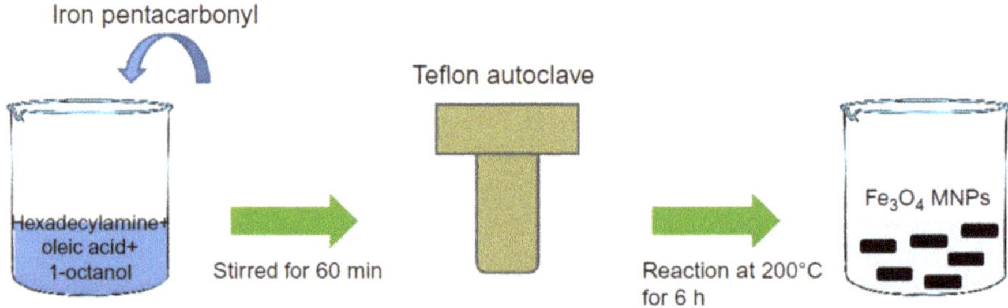

Figure 4. A schematic describing the hydrothermal synthesis approach. The process is similar to the co-precipitation; however, after preparing the mixture, the solution is transferred to an autoclave for further aging at high temperature and pressure for several hours [26].

2.4. Sol.-Gel and Polyol Methods

Sol-gel and polyol methods use essentially the same process but in a different direction, in which the sol-gel process is an oxidation reaction whereas the polyol process is a reduction reaction. The sol-gel synthetic approach is a well-known and widely used method in material science for the fabrication of metal oxides. This method usually starts with a colloidal solution acting as a precursor for either discrete nanoparticles or network polymers. Typically, a sol is a stable dispersion of colloidal nanoparticles or polymers in a solvent. Similarly, the gel could be either a colloidal gel, a network built from the agglomeration of colloidal nanoparticles, or a polymer gel, in which the nanoparticles have a polymeric sub-structure made by aggregation of sub-colloidal nanoparticles. Sol-gel processes usually are done at room temperature and the heat treatment can be done if high crystalline structures are desired [34]. The sol stage plays a critical role in the quality of the final nanoparticles made through this approach because the final size and saturation magnetization of the nanoparticles highly depends on the sol stage. The shape and crystallinity of the nanoparticles produced by this method are very sensitive to the type of precursors of the initial colloidal solution. As a result, this method is capable of producing nanospheres, hollow nanocages, and nanorods by controlling the water to acid ratio. Further adjustment on the temperature, pressure, and hydrous state can be done to tailor the phase of the nanoparticles [35].

In the polyol method, on the other hand, the polyols serve as both solvent and reduction agent and it applies stabilizers to prevent nanoparticles aggregation while controlling the growth of nanoparticles. The polyol method is done at high temperatures, the boiling temperature of the solution, but it does not require to be done at high pressure as it is required by the hydrothermal methods. This method can be done using different polyol solvents, for example, triethylene glycol (TREG), with high uniformity of morphology and colloidal stability nanoparticles. The magnetic nanoparticles produced by the sol-gel and polyol methods contain hydrophilic ligands on the surface that enhance their colloidal stability in the aqueous and non-aqueous solvent, an advantage compared to the magnetic nanoparticles produced by the co-precipitation method. Regardless of the high cost and safety issues associated with the sol-gel and polyol methods compared to the co-precipitation method, the sol-gel and polyol methods result in magnetic nanoparticles with significantly higher crystallinity and saturation magnetization [30].

2.5. Microemulsion Methods

Microemulsions are isotropic, stable, and clear mixtures of water, oil, and a surfactant [26]. The most commonly used microemulsion approaches for the synthesis of the magnetic nanoparticles are reverse, in which water dispersed in oil (w/o), and direct, in which oil dispersed in water (o/w) [21]. The surfactant could be a monolayer molecule with a hydrophilic tail dissolved in the water and a hydrophobic head dissolved in the oil, or vice versa.

Figure 5 schematically shows the microemulsion method, where the blue circles (also known micelles) are the magnetic nanoparticles precursors surrounded by surfactant molecules. The initial concentration and form of the surfactants are the keys to the final size and growth of the magnetic nanoparticles. For example, Darbandi et al. reported highly uniform size distribution and crystalline magnetic nanoparticles using the microemulsion method at room temperature [36]. It was shown that the presence of the surfactant residuals on the magnetic nanoparticles provides high molecular bonding affinity that makes this method highly desirable for producing magnetic nanoparticles for the detection and purification of the proteins in a solution as well as delivering vitamins [37].

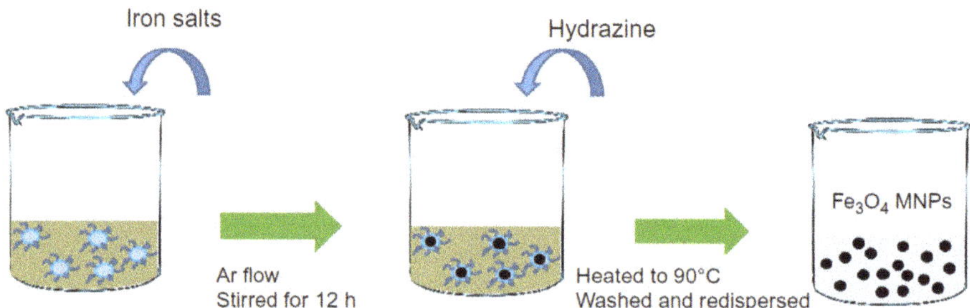

Figure 5. A schematic for the microemulsion process [26]. In this process, the iron (II) sulfate and iron (III) chloride salts are used in addition to hydrazine, which decreases the nanoparticles formation.

2.6. Sonolysis or Sonochemical Methods

Sonolysis or sonochemical methods employ high-intensity ultrasound irradiation to take advantage of the chemical effects induced by the acoustic cavitation for forming novel magnetic nanoparticle structures [38,39]. The ultrasonic irradiation creates bubbles that undergo continuous compression and expansion leading the oscillation of the bubbles (Figure 6). The oscillating bubbles accumulate the ultrasonic energy that continuously increases until causing the collapse and releasing the stored energy in the bubbles. Once the bubbles collapse, a highly localized energy burst is released that significantly increases the temperature and pressure at an extremely short time. In general, ultrasound-based irradiation is an excellent pathway for producing nanocomposites, such as dispersed magnetic nanoparticles in reduced graphene oxides or magnetic nanoparticle-loaded latex beads [40]. Even though the sonolysis or sonochemical method are promising for the fabrication of magnetic nanoparticles with desired sizes and excellent magnetic saturation properties, this method suffers from the dispersity and controllability of the magnetic nanoparticles' shapes. Furthermore, the magnetic nanoparticles synthesized using this technique are usually amorphous, porous, and agglomerated [41].

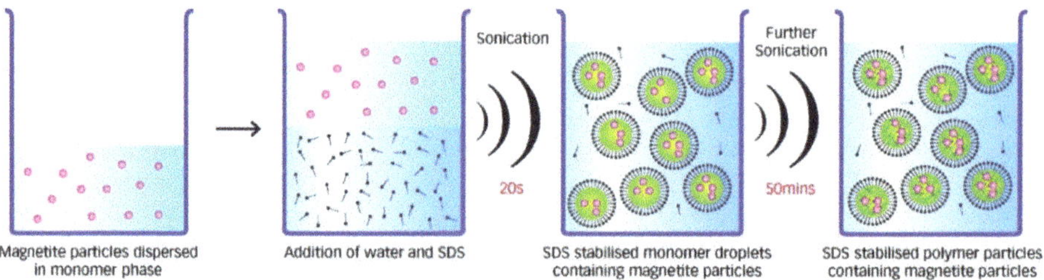

Figure 6. A schematic of the encapsulation of magnetic nanoparticles in latex nanoparticles using the sonochemically-driven miniemulsion polymerization technique [42]. Sodium dodecyl sulfate (SDS) is added to magnetite nanoparticles in the monomer phase followed by sonication to stabilize the surrounding magnetite nanoparticles.

2.7. Microwave-Assisted Synthesis

Microwave radiation forces molecules to reorient and oscillate with the electric field of the microwave signal (Figure 7). The strong oscillation at microwave frequencies results in intense internal heating that not only reduces the synthesis time but also significantly reduces the costs for nucleation and growth of the resulting magnetic nanoparticles [43,44]. The homogeneous excitation of the molecules using microwave signals have made this approach a strong tool for preparing the magnetic nanoparticles with controllable shape and size. One of the main advantages of the microwave-assisted synthesis is that this method can produce magnetic nanoparticles with different phases with an instantaneous coating that are desired for many applications, such as biomedical applications [30]. Furthermore, the microwave-assisted method is capable of producing magnetic nanoparticles with high colloidal stability that can be readily dispersed in water without any costly and complicated procedures for purification and ligand exchange [45]. These capabilities have made the microwave-assisted synthesis competitive to the thermal decomposition method for the mass production of magnetic nanoparticles. This method has been used for synthesizing magnetic nanoparticles as small as 6 nm up to 1000 nm and saturation magnetization comparable to the bulk values, where the crystallinity enhances with the increasing temperature of microwave heating [46].

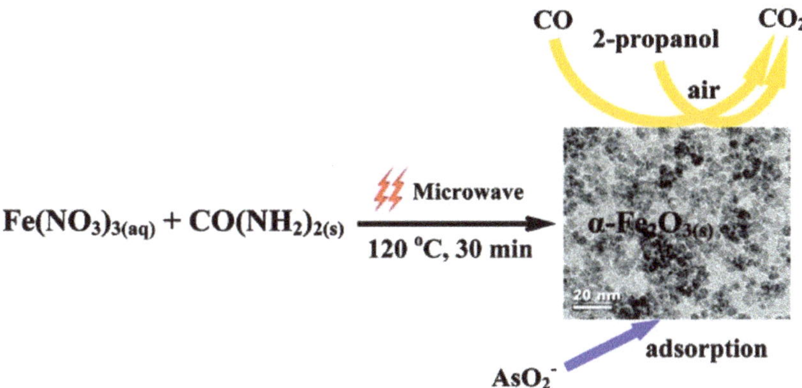

Figure 7. A schematic of magnetic nanoparticle preparation using the microwave-assisted technique at a temperature of 120 °C using a mixture of ferrite-nitrate and urea aqueous solution [47].

2.8. Electrochemical Deposition

The template-assisted electrochemical deposition has been broadly used in the synthesis of the magnetic nanowires as this method provides a highly controllable route for achieving precise dimension and compositions [48] (Figure 8). The template-assisted method can be divided into two categories depending on the template utilized, polymeric templates or anodic templates. The most commonly used polymer for synthesizing magnetic nanowires is polycarbonate because of its cost-effective approach, biocompatibility, and hydrophilic properties achieved by coating the polycarbonate templates. The hydrophilic property is the key for the fabrication of uniform and high aspect-ratio magnetic nanowires [49]. The polycarbonate templates are produced by ion irradiation of the row polycarbonate temples followed by a chemical etching process for opening the pores, which determines the final pore diameter. Due to the randomness of the ion irradiation, the distribution of the nanopores in polycarbonate templates is non-uniform. This features of polycarbonate templates have been used to synthesize interconnected networks of magnetic nanowires [50]. The aluminum anodic oxide templates, on the other hand, are relatively more expensive compared to the polycarbonate templates but they provide magnetic nanowires with very uniform diameters. The anodic aluminum oxides are prepared using both one-step anodization and two-step oxidation process after patterning an aluminum foil, where the two-step anodization leads to a significantly uniform distribution of the nanoporous [51,52]. A distinct advantage of anodic aluminum templates is that they provide flexibility to engineer the diameter along the porous leading to fabricate multi-diameters or tapper magnetic nanowires [53]. Particularly, electrochemical deposition is a strong tool for the synthesis of the magnetic nanowires as multi-segmented and/or multi-component, such as alloys.

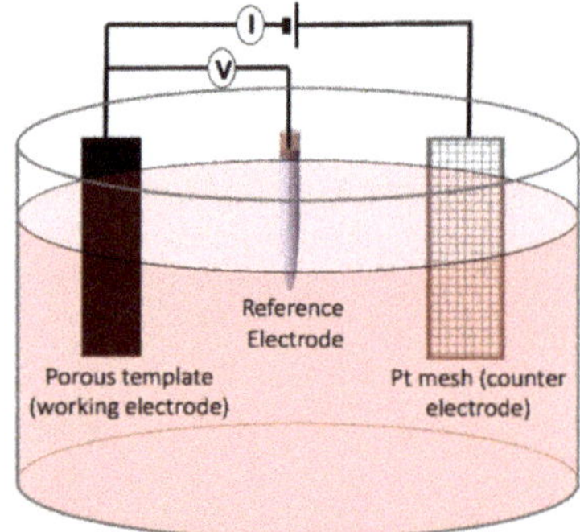

Figure 8. A schematic showing the synthesis of magnetic nanowires using a template-assisted electrodeposition technique [54]. After making the electrical contacts, the template is placed at anode and a platinum mesh is used as counter-electrode. The deposition current and voltage are controlled using the reference electrode, where the electric field between the working and counter electrodes forces the free ions to be deposited in the pores.

It should be mentioned that electrochemical deposition has been also used for synthesizing non-cylindrical magnetic nanoparticles, such as nanospheres [55]. Despite the high costs associated with the electrochemical synthetic approach, this method is very useful compared to other methods if very high purity products with controlled size and shape are demanded. The electric field distribution within the electrodes plays a critical role in the magnetic nanoparticle size and aggregation during the synthesis. A challenge during the synthesis of magnetic nanoparticles using the electrochemical methods is the formation of the metallic Fe. To overcome this challenge, one should apply a more anodic potential over a long time, which makes this method unfavorable compared to the other synthetic methods [56] for mass-production.

2.9. Biosynthesis Methods

The biosynthesis method employs a microbial enzyme or a plant phytochemical with reducing properties that make this method eco-friendly. Traditionally, magnetotactic bacteria and iron-reducing bacteria are used to synthesize magnetic nanoparticles [57]. This method was shown promise for doping iron oxide magnetic nanoparticles with cobalt with improved magnetization properties [58]. More recently, several attempts have been put forth to employ human stem cells to synthesize re-magnetized magnetic nanoparticles [59]. The type of the bacteria and the synthesis conditions (aerobic or anaerobic), determine the phase of the resulting magnetic nanoparticles. For example, maghemite magnetic nanoparticles with superparamagnetic characteristics can be produced using *Actinobacter* bacteria under aerobic conditions [60]. In general, the mechanism of the biosynthesis of magnetic nanoparticles has not been well understood in order to clarify the controlling parameters of the shape and sizes while maintaining the desired saturation magnetization of the magnetic nanoparticles [61].

2.10. Other Techniques

In addition to the aforementioned syntheses that are chemical approaches, magnetic nanoparticles have been also produced using physical methods. The physical techniques are in their majority top-down processes where a bulk magnetic material is decomposed into magnetic nanoparticles. A few well-known examples of physical techniques are mechanical milling [26], electrical explosion of wires [62], and laser target evaporation [63]. Even though the physical approaches have higher production yield, up to 200 g/hr, they form almost 10% of magnetic nanoparticles for diverse applications.

This is because the physical approaches have relatively high power consumption and controlling the shape and size distributions is difficult.

3. Different Shapes

3.1. Sphere, Cubic, Octahedral, and Rhombohedral

The final morphology of the magnetic nanoparticles not only depends on the synthetic method used, but also on the solution composition, temperature, and pressure [30]. For example, the precipitation of an iron (II) salt in alkaline media in the presence of a mild oxidant such as potassium nitrate results in magnetic nanocubes [64] rather than nanospheres. This is because the aforementioned parameters accelerate crystal growth over specific facets while hindering the growth over other facets by slowing it down. Aside from the aforementioned parameters, the ratio of the compounds in the solution is the most critical parameter determining the shape of the magnetic nanoparticles. For example, when the Fe (II):OH$^-$ ratio gets close to 0.77, the achieved magnetic nanoparticles are cubic, while the shape changes to octahedral as the ratio approaches 1.65. It should be noted that the crystal growth at different facets is also controllable by precise adjusting the pH during the synthesis. At low pH, the OH$^-$ concentration is negligible thus the growth takes place mainly by aggregation, where the growth kinetics are much faster. In another words, the primary nanoparticles are not repelling each other because they are not sufficiently charged and the aggregation is followed by subsequent recrystallization leading to octahedral nanoparticles with a broad range of sizes, from a few nanometers to micrometers [65,66].

As mentioned, thermal decomposition is able to differentiate the nucleation and growth process during the synthesis. By changing the precursors, it is possible to obtain octahedral magnetic nanoparticles using a thermal decomposition method. For example, the decomposition of iron (II) oleate in tetracosane in the presence of oleylamine results in octahedral nanoparticles nucleated by the selective binding of oleylamine to {111} facets [67]. Interestingly, the heating rate of the solution also was found to lead to octahedral nanoparticles. A synthetic strategy laying between the aqueous and organic media consists of the hydrolysis of Fe (II) acetate in the presence of oleylamine dissolved in xylene. Heating the reaction mixture followed by a fast injection of water triggers the hydrolysis of the Fe-oleylamine complex leading to nanocubes as well.

Recently, a three-step approach was reported for the synthesis of the rhombohedral magnetic nanoparticles [68] through a three-step synthetic approach: (1) synthesis of antiferromagnetic nanoparticles, (2) nanoparticle coating, and (3) subsequent reduction of the core material to magnetite. This approach comprises the generation of hematite nanoparticles, further encapsulation in silica and final reduction to magnetite, in which the first step is crucial for the growth of rhombohedral scaffolding and it is achievable by solvothermal synthesis.

3.2. Nanodiscs

The easiest synthetic approach for the fabrication of the nanodiscs is the precise control of the deposited charge during the electrochemical deposition. Since the length, or thickness in the case of nanodiscs, can be accurately controlled using the applied potential and deposition time, the electrochemical deposition is the strongest tool for achieving this morphology. However, if nanodiscs of a very small size are desired, this technique becomes costly compared to the other synthetic approaches, even for small volumes of magnetic nanoparticles. Alternatively, both one-step and two-step synthetic approaches, such as the solvothermal technique, become a good alternative [69]. For the one-step synthetic techniques, a promising approach must delay the nucleation of the nanoparticles. The nucleation can be delayed in the absence of the water during the formation of the common nanoparticles, such as nanospheres [65]. In the aforementioned aquatic-based synthesis techniques, water acts as an accelerating agent, which increases of the thickness while suppressing the diameter growth. Reducing the size of nanodiscs while maintaining the aspect ratio is also possible through two-step synthesis approaches [70]. Moreover, it was reported that hematite nanodiscs can be achieved by the hydrolysis of iron (II) chloride in a mixture of water/ethanol in the presence of sodium acetate [71]. In this approach, the diameter and thickness of the hematite nanoparticles can be controlled by the amount of water in the solvent and sodium acetate.

3.3. Elongated Nanoparticles

The elongated nanoparticles are the results of hindering the crystal growth rate at a specific direction while accelerating the growth rate at other facets. This can be done by incorporating precursors during the synthesis of the template-free synthetic approaches or simply using a template during the synthesis. The elongated nanoparticles are technically the 1D magnetic nanoparticles which are described in the literature with different names such as nanowhiskers [72] or nanorods [73], nanorices [74], nanobelts [75], nanospindles [76], and nanowires [77]. The different terms are attributed to their final morphology, the geometry of the edges, and the axial ratios of the lateral dimension to the longitudinal dimension. Recently, numerous attempts were taken place to achieve a one-step direct synthesis of elongated nanoparticles as opposed to the conventional methods as they require templates for controlling the morphology. Comparatively, the final properties, such as crystallinity, of the elongated nanoparticles highly depend on the synthetic approach used [65]. The template-free synthetic approaches are mainly limited to magnetic nanoparticles with small aspect-ratios [78], on the order of 3–10. For example, the hydrolysis and oxidation of the iron (II) sulphate in water in the presence of carbonate ions was shown to result in magnetic nanoparticles with an aspect ratio of 3 to 4 at room temperature [74]. In this approach, the concentration of the iron (II) sulphate and carbonate, air flow rate and reaction time, additive precursors, and pressure are all influencing factors.

The strongest synthetic approach for elongated nanoparticles, especially with a broad range of aspect ratios in order of 1 to 1×10^4, is template-assisted electrochemical deposition [77]. By adjusting the potential and deposition time, the aspect ratio can be easily controlled. The final aspect ratio depends on the ratio of the template pore diameter to thickness. As mentioned, the main advantage of the electrochemical deposition is that it can synthesize multi-compounds and multi-segmented elongated nanoparticles. The length uniformity can be controlled by applying step potentials, additive precursors, and the solution temperature.

4. Characterization

Similar to other types of nanoparticle, precisely characterizing of the magnetic nanoparticles is very important as it determines the reproducibility of the results. Characterization of magnetic nanoparticles in terms of their size, shape and composition is particularly substantial as their magnetic properties are significantly influenced by those parameters. For example, if the size of a magnetic nanoparticle shrinks sufficiently such that it no longer can hold a domain wall, the coherence between the spins results to superb magnetic properties. Similarly, the shape of the magnetic nanoparticles causes magnetic inhomogeneity leading to quantum effects that cannot be achieved in the bulk states. In nanobiotechnologies, particularly in biosensing, the accurate characterization of the composition and surface coating is vital because their biocompatibility and sensitivity are highly relied on these characterizations. In this section, we briefly explain the highly characterization techniques that have been utilized to demonstrate their functionalities in diverse applications, from quantum storage to nanomedicine.

4.1. Shape, Size, and Composition

4.1.1. Transmission Electron Microscopy (TEM)

TEM is a microscopy technique that exploits the interaction between a uniform flux of electrons and the nanoparticles under study. The interaction of electron flux leads to a part of it being transmitted through the nanoparticles while the rest are scattered, where the interaction depends on the size, shape, and elemental composition of the nanoparticles. TEM is the most popular technique to measure nanoparticles' shape, size, and homogeneity because it provides direct images of the nanoparticles. Recent advances in TEM imaging, such as liquid-phase TEM, not only directly characterizes magnetic nanoparticles size and morphology but also characterizes the interparticle distances in a solution [79,80], which was shown to be a critical parameter for the magnetic response of the magnetic nanoparticles. Interestingly, TEM facilitates the real-time imaging for demonstrating the dynamic transformation in nanoparticles over time [81]. For example, TEM was used to visualize the biodegradation of the coating of nanoparticles in the presence of biological entities, such as bacteria. It also was found to be a very strong approach for real-time monitoring of the dynamic growth of the magnetic nanoparticles in suspension [82]. Nevertheless, TEM is very costly and slow for the characterization of nanoparticle assemblies if there are a large number of them with polydispersity. Furthermore, due to high absorption of the electron energy with liquid molecules, this technique is very tedious to measure the size and shape of the magnetic nanoparticles in suspensions. In these cases, the other methods, such as nanoparticle tracking analyzer (NTA) or dynamic light scattering (DLS), are more efficient [83].

4.1.2. Dynamic Light Scattering (DLS)

Dynamic light scattering (DLS) is a commonly used technique to find the size of particles suspended in colloidal solutions (Figure 9). DLS uses the Stoke-Einstein law to relate the light scattered from the nanoparticles in a colloidal solution to their hydrodynamic diameter. It is beneficial to have solutions with low concentration to avoid simultaneous multiple scattering events in order to achieve a more accurate analysis. The DLS technique has been used to realize the influence of the nanoparticle shapes, size, concentration, and surface coating on the colloidal stability of the magnetic nanoparticles [83]. For middle-size magnetic nanoparticles, the DLS technique provides accurate results for determining the size that was shown to match with the results derived from TEM and SEM images. However, for small size magnetic nanoparticles, the DLS technique does not match the TEM results due to the radius of curvature effects. The DLS is not a good technique for analyzing the magnetic nanoparticles if the heterogeneity and poly-disparities are high [84]. That is because the larger nanoparticles scatter substantially more light obscuring the detection of the scattered lights from the small nanoparticles. Furthermore, the DLS requires transformative analysis with several assumptions, especially when the nanoparticles are non-spherical or the polydisparity is high, which diminishes its accuracy [81].

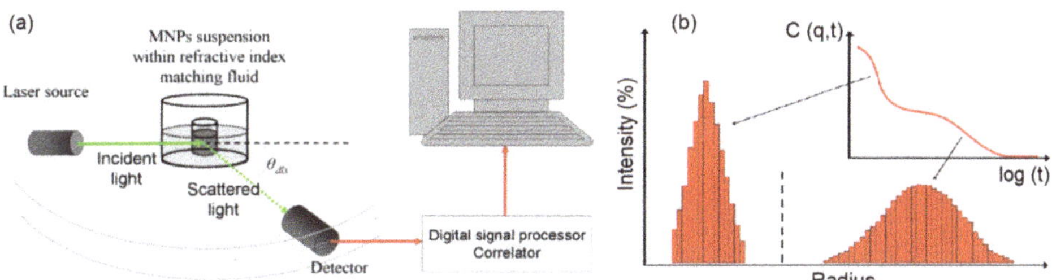

Figure 9. A schematic of the dynamic light scattering (DLS) setup [84] (**a**) and the acquired results (**b**). A laser light hits the magnetic nanoparticles in a continuous flow while the scatted light is monitored using a detector. The results are size distributions where the accumulative function can be calculated by taking an integral from the size distribution.

4.1.3. Nanoparticles Tracking Analyzer (NTA)

NTA takes advantage of both the light scattering and Brownian motion properties of the nanoparticles in a solution to determine the size distribution at a lower concentration limit compared to the DLS technique [85] (Figure 10). The main advantage of the NTA over DLS is that its results are not biased by the aggregation of larger nanoparticles. As a result, the literature shows that NTA provides more accurate results for both monodisperse and polydisperse samples compared to DLS [86].

The main difference between the NTA and DLS is that NTA tracks single nanoparticles while DLS analyzes an ensemble of nanoparticles with a high bias towards the larger nanoparticles. This different operation mode causes the NTA to be relatively slower with a more complex operation procedure than DLS. However, since NTA is also capable of preedictingt the magnetic nanoparticle concentrations [87], it has received a huge amount of attention over the last few years.

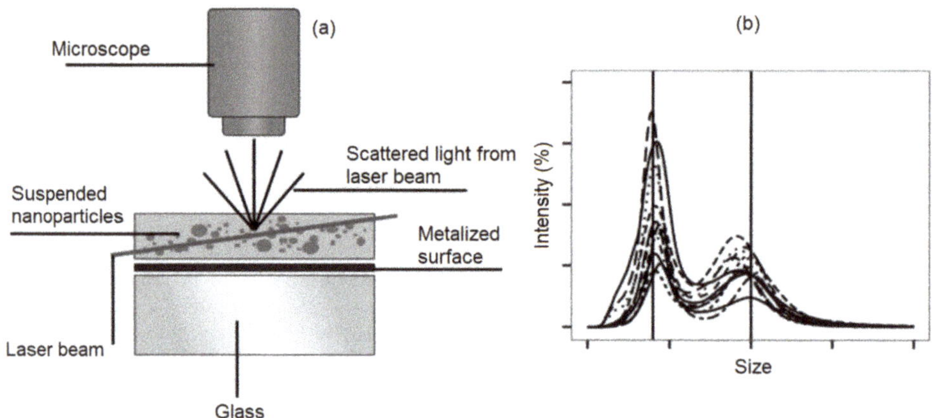

Figure 10. A schematic of the nanoparticles tracking analyzer [85]. (**a**) NTA employs a high speed camera to capture the light scattered by the Brownian motion of nanoparticles to determine the nanoparticles size distributions, and (**b**) is an example for a sample including two size distributions.

4.1.4. X-ray Diffraction

Undoubtedly, X-ray diffraction is one of the key tools that has been extensively used to characterize both magnetic and non-magnetic nanoparticles. The X-ray diffraction technique provides detailed information regarding the phases, lattice parameters, crystalline grain size, and crystalline structures [81]. The composition, crystal structures and the nature of the phases can be determined by comparing the position and intensity of the X-ray peaks with the available reference database while the crystalline grain size can be determined using the broadening of the peaks [88]. Aside from the amorphous nanoparticles, the broadening of the X-ray peaks is mainly due to the nanoparticle/crystalline size and lattice strains. Practically, if the magnetic nanoparticles are big enough to hold more than one crystal boundary, the X-ray cannot distinguish between the boundaries leading in the misrepresentation of the crystalline grain sizes [89].

Numerous modifications have been applied to the X-ray diffraction to enhance its capability beyond demonstrating the chemical composition and/or crystallinity. For example, X-ray absorption includes both extended absorption fine structure and X-ray absorption near edge structure capable to measure the surface binding energy and density of states [90]. Another example is X-ray photoelectron spectroscopy that has been widely used for surface chemical analysis, electronic structures, elemental composition and oxidation state of the elements [90].

4.1.5. Fourier Transform Infrared Spectroscopy

Fourier transform infrared spectroscopy (FTIR) is another technique for determining the structure, size, and composition of magnetic nanoparticles. FTIR measures the absorption of electromagnetic radiation with wavelengths within the mid-infrared region. By comparing the FTIR spectra of pure magnetic nanoparticles or their modified versions, such as after adding a surface coating, one can quantify the composition (Figure 11). Once a molecule is excited with IR radiation, its dipole moment gets aligned and oscillates accordingly. Thus, the recorded spectrum gives the position of the bands related to the strength and the nature of the bonds providing information regarding the molecular structure and inter-molecule interactions [91]. FTIR is usually combined with differential electrochemical mass spectroscopy to further detect the volatile reactants. For example, Shukla et al. employed this method to illustrate the surfactant bonding on iron platinum (FePt) nanoparticles stabilized in a non-polar solution [92]. They showed that FTIR is capable of detecting the types of the bonding, either monodentate or bidentate, on the FePt magnetic nanoparticles. The relative low cost and high throughput of this technique have resulted in several attempts to push its limit to analyze magnetic nanoparticles of a few nanometers in size. For example, it was shown that FTIR can determine the crystallinity of magnetic nanoparticles with the average size below 15 nm [93].

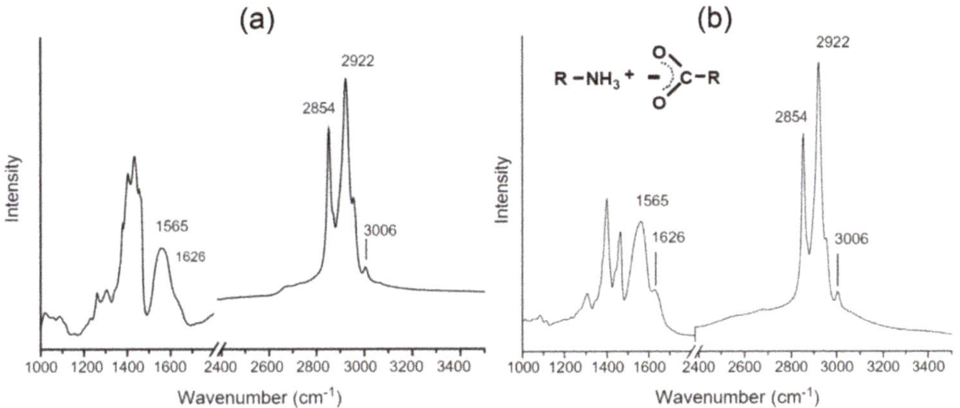

Figure 11. Comparing the FTIR spectra of (**a**) modified magnetic nanoparticles, (**b**) the modification composition [92].

4.1.6. Nuclear Magnetic Resonance (NMR) Spectroscopy

NMR spectroscopy is an analytical method for the quantitative determination of magnetic nanoparticle structures. NMR spectroscopy is based on magnetic resonance of the nanoparticles' nucleii under a strong magnetic field that induces an energy difference between the down and up spins. Even though NMR spectroscopy is a useful method for studying superparamagnetic nanoparticles, it fails to characterize ferromagnetic nanoparticles, such as nickel or cobalt magnetic nanowires, because their strong magnetization can cause a reduction of the relaxation time while drastically shifting the signal frequency and local magnetic field [94]. The NMR method has been used for characterizing the surface coating of the magnetic nanoparticle as it is very sensitive to electronic structures and molecular bonding on the surfaces [95] if they are sufficient to be detected. Practically, the NMR technique has sensitivity on the order of a few nanogram that can be enhanced to a few picograms [96].

4.1.7. Mass Spectroscopy (MS)

Mass spectroscopy (MS) is known as a powerful and reliable tool for the analytical characterization of magnetic nanoparticles. The common MS tools have a sensitivity that is on the order of a few picograms, which is better than the conventional NMR technique [97]. This technique not only provides elemental information regarding the composition, chemical states, and structures of the magnetic nanoparticles but also quantifies the surface bioconjugations for targeting biomolecules [81]. Aside from the simplicity and universality of MS, it is a highly sensitive technique that can be coupled with separation techniques for real-time sorting applications. For example, the inductively coupled plasma MS (ICP-MS) is a robust, highly sensitive technique with a wide dynamic range for elemental analysis of the magnetic nanoparticles [98]. Another example is the single-particle operation mode ICP-MS those improvements on the identification of the concentration and size distribution of magnetic nanoparticles [99].

4.1.8. Thermal Gravimetric Analysis (TGA)

Even though FTIR provides information regarding the presence of binding on the surface and its type, it does not provide volumetric information, such as mass to mass ratio of the magnetic nanoparticles to its coating. This limitation can be addressed using thermal gravimetric analysis (TGA), which provides information about the composition and mass of the coating [100–102]. This technique increases the sample temperature while monitoring the mass change of the sample as it changes due to the degradation of components. As a result, this technique is not only capable to quantitatively determine the mass of the coating of the magnetic nanoparticles but also determines the compositional purity and the thermal stability of the coating [103]. For example, Ziegler-Borowska et al. conducted TGA experiments in both air and nitrogen where they showed the modified chitosan has low thermal stability; however, once this coating thermally degrades, its degradation products form a stable surface layer [104]. The drawback of this method is its destructive analysis and the minimum initial nanoparticle mass [81].

4.2. Magnetic Characterization

4.2.1. Hysteresis Loop Measurement

As the size of the magnetic materials decreases towards the nanoscale, they exhibit substantially different magnetic properties compared to their bulk state. This change in the magnetic properties is due to two facts. First, the magnetic nanoparticles can no longer hold multiple domains to balance their magnetiostatic energy and exchange energy [105]. Second, the number of the atoms on the surface becomes comparable to the number of atoms in the volume. The former means that they behave as a single domain nanoparticle, where they behave like superparamagnetic if the thermal fluctuations become significant compared to the energy barrier. The latter means that, at a few nanometer size, the unpaired electrons become more dominant leading to additional anisotropies, such as surface anisotropy. The major hysteresis loop measurement is the key technique to measure the basic hysteretic information of any magnetic nanoparticle. The hysteresis loops can be measured using a superconducting quantum interface device (SQUID) or vibrating sample magnetometry (VSM), in which the SQUID has a significantly higher resolution. The major hysteresis loop provides the saturation magnetization, remanence magnetization, and the coercivity. These parameters are sufficient to describe the magnetic response of isolated-single domain magnetic nanoparticles, or those with negligible magnetic interactions, with respect to a magnetic field. However, if further characterization is required or the interaction fields among the magnetic nanoparticles are not negligible, more advanced magnetic characterization techniques are needed [106–108]. Importantly, the hysteresis loops are unable to discriminate between the different phases if there are more than one in magnetic nanoparticle assemblies.

4.2.2. Mössbauer Spectroscopy

Mössbauer spectroscopy (Figure 12), is a powerful analytical technique to evaluate the oxidation state, spin states, and spin ordering of Mössbauer-active elements (such as Fe) for the identification of magnetic phases [109,110]. Mössbauer spectroscopy can also provide insight into the quantification of the thermal blocking/unblocking (superparamagnetic response) and magnetic anisotropy energy if the measurements are conducted as a function of temperature. The Mössbauer spectroscopy shift is an important parameter that arises from the nuclear-energy shift that is caused by the Coulombic interaction between the nucleus and the electron density at the site of the nucleus [81]. As an example, Oh et al. used Mössbauer spectra as a probe to quantitatively study the local state of electrons on the surface of the FeCo nanoparticles [111]. They investigated the process parameters on the magnetic properties of the FeCo nanoparticles using Mössbauer spectroscopy. Furthermore, Lange and et al. showed that Mössbauer spectra can be used to determine the hyperfine interactions between nuclei and their surroundings because this method is very sensitive to the local structural and chemical environment of the Mössbauer-active elements [112]. In this direction, Tiano and co-workers utilized Mössbauer spectroscopy and SQUID to draw the relation between the magnetic properties and the composition of several magnetic nanoparticles, such as Mg, Fe, Co, and Ni [113]. The Mössbauer isomer shift is not technically a probe for determining the oxidation number of the dopant atoms because it only determines the charge state on the nucleus [114]. Therefore, if both Fe^{2+} and Fe^{3+} species persist in a Fe-doped nanoparticles, the electron redistribution from the dopant sites to the crystal matrix leads a very similar shifts for both species.

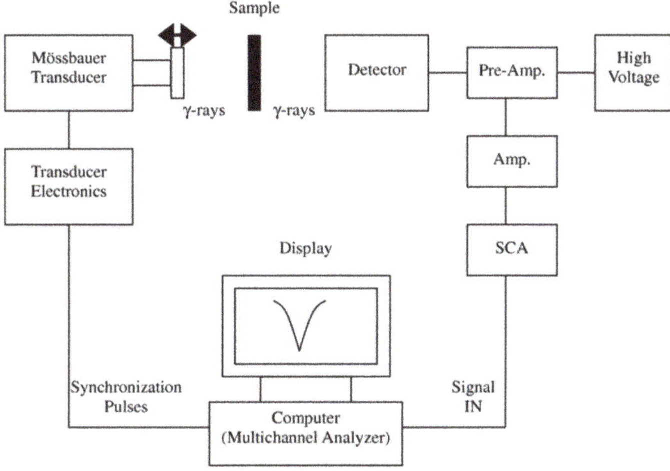

Figure 12. A schematic of the Mossbauer spectroscopy setup [111].

4.2.3. Ferromagnetic Resonance (FMR)

FMR is a spectroscopic technique probing the nanoparticle magnetization induced by the magnetic moments of the dipolar coupled of unpaired electrons. The width of the FMR peak was shown to be related to the size, shape, defects density, and surface anisotropy. For example, Diehl et al. conducted FMR spectroscopy experiments on both crystalline and imperfect magnetic nanoparticles revealing that this technique is able to illustrate the coherence of the lattice drastically impacts the magnetic inhomogeneity and anisotropic properties of nanoparticles [115]. The same authors also reported that surface defects can be detected at low temperatures using FMR. It was shown that decreasing the temperature or increasing the magnetic nanoparticle size causes a shift in the resonance field leading to enhancement of the FMR peak asymmetry and broadening [115]. It should be mentioned that FMR usually predicts the size of the nanoparticles smaller than the actual values measured by TEM due to the presence of the magnetically disordered layer on the surface.

4.2.4. Magnetic Susceptibility

Magnetic susceptibility is defined as the ratio of the magnetization to the applied field indicating how strongly whether a nanoparticle is repelled by or attracted into a magnetic field. This technique is able to quantify the magnetic nanoparticles coated with polymeric or organic/inorganic materials. Consequently, this technique has been used to determine the mass of the surface coatings to the magnetic mass of nanoparticles [116,117]. Magnetic susceptibility also has been used to acquire the size and size distribution of magnetic nanoparticles, and a good agreement was observed by comparing to the TEM results [117]. The field-dependent magnetic susceptibility was shown to provide information regarding the magnetocrystalline anisotropy of the magnetic nanoparticles. Due to the limited dynamic range of the susceptometers, this technique most often used over a range of temperature rather than frequencies that provides insight into the heating efficiency of the magnetic nanoparticles for certain applications.

4.2.5. Electron Holography

Electron holography is a holographic technique using electron waves for imaging. This technique uses a high spatial and temporal coherence of an electron beam to acquire holographic imagery. The main advantage of electron holography is that this technique provides direct information regarding the internal magnetic structures rather than via the stray fields or only the surface magnetization states (Figure 13). Among the several different holography configurations, the off-axis and in-line configurations are the most commonly used techniques in imaging nanoparticles. For example, Ortega et al. reported the off-axis electron holography under Lorentz microscopy conditions to observe the magnetization distribution and to determine the saturation magnetization of multi-segmented FeGa/Cu nanowires [118]. They showed the presence of an antiferromagnetic configuration along the FeGa/Cu nanowires even though the magnetic nanowires are ferromagnetically coupled. Electron holography was also shown to be useful for visualizing the magnetic domain walls and/or domain wall pinning in magnetic nanoparticles with high resolution compared to other magnetic imaging techniques, such as magnetic force microscopy. For example, Biziere et al. reported on observation of the magnetic domain walls in magnetic nanowires leading to illustrating the presence of two types of domain walls transferring to each other by manipulating the diameter [119].

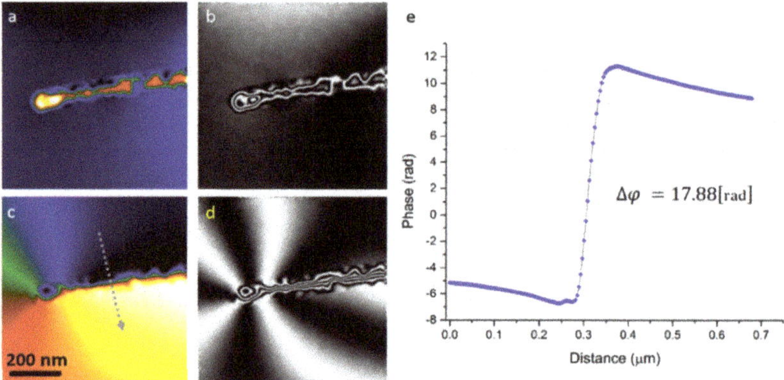

Figure 13. An exemplary experimental result showing magnetic nanowires captured using the electron holography technique [120]. (**a,b**) are the electrostatic phase and its amplification, respectively. (**c,d**) are the magnetic phase and its amplification. (**e**) is the phase shift across the arrow in (**c**).

4.2.6. Remanence Curves Technique

The remanence curves technique has been widely used to qualitatively determine the nature of the interaction fields among magnetic nanoparticles in assemblies. The remanence curves technique characterizes magnetic nanoparticles by subtracting isothermal remanence and DC demagnetization curves and comparing the results with the Stoner-Wohlfarth model, also known as Henkel plots. The isothermal remanence measures the remanence by applying and removing an ascending field to initially demagnetized nanoparticles [121], while the DC demagnetization measures the remanence by applying and removing a descending field to initially saturated nanoparticles [122]. For non-interacting nanoparticles, both curves are identical, meaning there is no interaction. However, the curves deviate for interacting nanoparticles, where the sign and strength of the deviation qualitatively determine the interaction fields. Recently, several attempts have been made to use this technique for quantitative analysis of the interaction fields [123,124]. For example, Huerta et al. used it to study the interaction fields between arrays of magnetic nanowires where the authors showed interaction fields are dominated by dipole-fluctuation effects [123]. In their approach, they used the difference between the field where the DC demagnetization curve is zero and isothermal remanence curve is 0.5 to quantify the interaction fields. In another example, Moya et al. used the remanence curves technique combined with the ZFC/FC to quantify the interaction fields [125].

This technique also has been proposed to probe the spin disorder in magnetic nanoparticles [126]. For example, Toro et al. reported that the dips in the DeltaM plots, calculated by subtracting the remanence curves, are not necessarily due to the interaction fields as they could be due to spin disorder, an inhomogeneity of spin distributions in the magnetic nanoparticles [127]. Despite the simplicity and straightforward measurements and analysis of the remanence curves technique, this technique is limited to magnetic nanoparticles having strong magnetic anisotropy, in which their coercivity is larger than the interaction fields [128]. As a result, it is unable to determine the interaction fields among the superparamagnetic nanoparticles, or even the ferromagnetic nanoparticles if the interaction fields are comparable to the coercivity. An interesting comparative analysis on this was given by Zamani Kouhpanji et al., where the authors compared the results of the hysteresis loops, remanence curves, first-order reversal curves (FORC), and projection method to assess the reliability and validity range these techniques for magnetic characterizing of nanoparticles [107].

4.2.7. Magneto-Optic Kerr Effect (MOKE) Microscopy

MOKE microscopy is an optical imaging technique that utilizes the interaction between the magnetization and optical waves to image the magnetization state on the surface of nanostructures (Figure 14). The MOKE measurements are categorized into four groups depending on the relative direction of the light polarization and the magnetization direction. The MOKE technique has been used to study the angular dependence of the magnetization demonstrating the anisotropy and the reversal mechanism of the magnetic nanoparticles [129]. For example, Palmero et al. reported on the coercivity mechanisms in multi-diameter magnetic nanowires illustrating the fundamentals for controlling the propagation of the single domain wall in the magnetic nanowires [130]. In another example, Bran et al. reported on capturing the Barkhausen jumps inside the magnetic/non-magnetic multi-segmented nanowires using the MOKE technique [131]. They showed that the magnetization can be pinned at the interface between the magnetic and non-magnetic segments that can be engineering using the ratio of the segments.

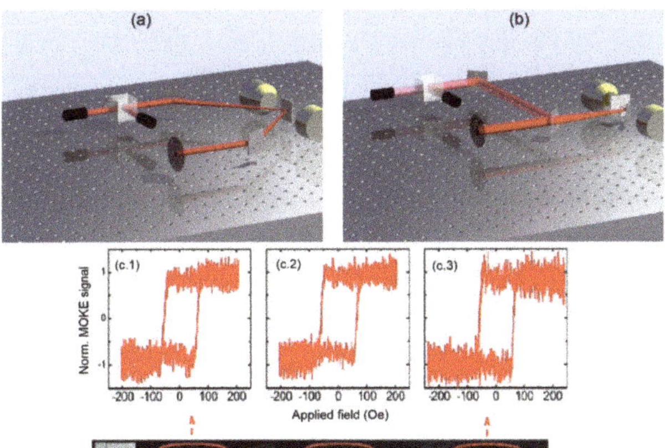

Figure 14. Schematic illustrating the magneto-optic Kerr effect (MOKE) setup [129]. (**a**) Out-of-plane measurement and (**b**) in-plane measurement, (**c.1–c.3**) show the MOKE signals at different locations along the magnetic nanowire corresponding to the red ellipses [132].

4.2.8. Magnetic Force Microscopy (MFM)

The MFM technique is a form of scanning probe microscopy that measures the magnetic forces applied on the probe for imaging. In the MFM, a nanoscale probe is coated with a few tens of nanometers of magnetic material to capture the magnetic stray fields of the magnetic nanoparticles. The main advantage of the MFM is that it can image the magnetic domains indicating the magnetization direction. As an example, Mohammed reported on controlling the spin-torque driven domain wall motion in staggered magnetic nanowires using MFM [133]. Furthermore, it has been widely used to indicate the interaction fields among the magnetic nanoparticles [134]. Because of the high sensitivity of this technique, it has also been used to quantify the amount of magnetic nanoparticle internalized into the cells where optical imaging, such as fluorescence imaging is not feasible [135]. The MFM technique was originally proposed to measure the remanence magnetization in the absence of an external field, in other words, it used to only measure the remanence magnetization. However, recently several attempts were taken that enabled the whole hysteresis loop measurement using MFM in the presence of an external field. For example, Coisson et al. reported on implementing the MFM to probe the magnetic vortex chirality of magnetic nanodisc using the MFM technique [136].

4.2.9. Minor Hysteresis Loops

Minor hysteresis loop measurements are performed at fields smaller than the saturation field of the magnetic nanoparticles. These measurements are very useful to describe the magnetization response of the magnetic nanoparticles that do not experience full saturation during the implementation of a cyclic magnetic field, such as during hyperthermia or nanowarming experiments. For example, Shore et al. conducted minor hysteresis loop measurements to demonstrate the heating efficiency of the magnetic nanowires for nanowarming applications [137]. Since the minor hysteresis loops are measured at not full saturation, the overall shape of the hysteresis loops highly depends on the initial magnetization state of the nanoparticle. As a result, the minor hysteresis loops technique has been used to study the reversal mechanism not only in magnetic nanoparticles but also in complex magnetic nanostructures.

4.2.10. Zero-Field-Cooled/Field-Cooled (ZFC/FC) Magnetization Curves

The ZFC/FC technique measures the magnetization at different temperatures to determine the transmission from ferri/ferromagnetic response to superparamagnetic response [138]. In ZFC, the nanoparticles are cooled to the lowest temperature in the absence of an applied field. Then, the magnetization is recorded at a finite field while the temperature increases (Figure 15).

Initially, the magnetization increases with the temperature before it starts decreasing at a certain temperature, which is called the blocking temperature. The FC part of the measurement records the magnetization as the temperature is returned to the initial value. The blocking temperature has been widely used to determine the anisotropy of the magnetic nanoparticles. For example, Nemati et al. investigated the shape and size effects of nanospherical and nanocubic nanoparticles on their magnetic anisotropy and heating efficiency using the ZFC/FC technique [29]. In another example, Das et al. used the ZFC/FC technique as a probe for illustrating the structural transition of magnetic nanorods [139].

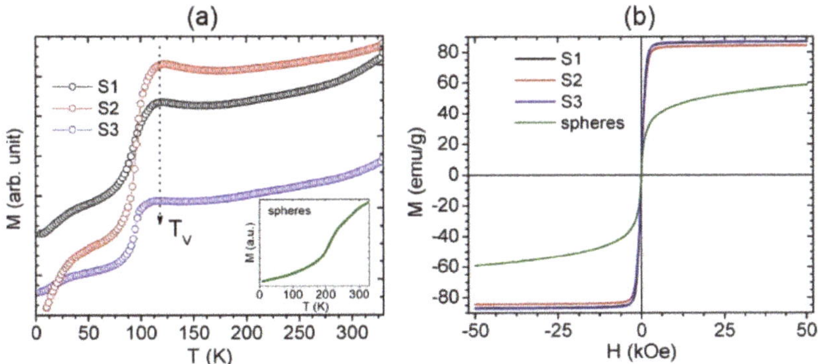

Figure 15. An example of the zero-field cooling (ZFC) measurement for three different types of superparamagnetic magnetic nanoparticles (**a**) and their corresponding hysteresis loops (**b**). As the temperature varies, there is a jump in the magnetization indicating the block temperature [140].

4.2.11. Angular-Dependent Coercivity

The angular-dependent coercivity technique measures the hysteresis loop at different angles between the applied field and the easy axis of the magnetic nanowires. Intuitively, it is only applicable to magnetic nanoparticles having a strong shape anisotropy, such as magnetic nanowires or nanorods. The variation of the coercivity in terms of the angle determines the reversal mechanism in the nanoparticles, which can occur via a coherent rotation or a domain wall propagation [105]. It was shown that the reversal mechanism can changes from the coherent rotation to a domain wall at some angles, where there is a sharp reduction in the coercivity after an initial increase [141]. This technique also has been used to determine the energy barrier and magnetic anisotropy in nanoparticles [142]. In general, since this technique is unable to directly measure the interaction field and subtract its effects from the coercivity, this technique is not well-suited for characterizing the magnetic nanoparticles when the interaction fields are not negligible [143].

4.2.12. First-Order Reversal Curve (FORC)

The first-order reversal curve (FORC) technique is one of the most powerful techniques that has been widely utilized for quantitative and qualitative description of complex magnetic nanostructures [144]. Figure 16 provides an example of the FORC data and heat-maps for analyzing the types and volume ratios of magnetic nanowires combination.

The FORC technique scans the whole area of the hysteresis loop leading to collecting detailed information regarding the interaction fields and intrinsic magnetic properties of the magnetic nanoparticles [145]. For example, Ramazani et al. utilized this technique to quantify the interaction fields and coercivity distribution of magnetic nanowire arrays [146]. In another example, Zamani Kouhpanji reported on the application of the FORC technique for demultiplexing the magnetic nanowires embedded inside the biological species [143]. Regardless of the complex data processing of the FORC [147–149], this technique is significantly slower than the other magnetic characterization techniques because it requires scanning the whole area of the hysteresis loop to a detailed analysis [107,108,129]. In this direction, several works have been done to enhance the measurements by introducing modifications to the standard protocol of the FORC technique. In a significant technological development, Zamani Kouhpanji et al. introduced a novel modification into the FORC protocol indicating only a narrow region of the hysteresis loop is required to fully characterize the magnetic nanoparticle assemblies [108]. They showed the interaction fields and coercivity of the magnetic nanoparticles can be comprehensively and reliably measured by a factor of at least 50×–100× faster than the standard FORC protocol [54,107].

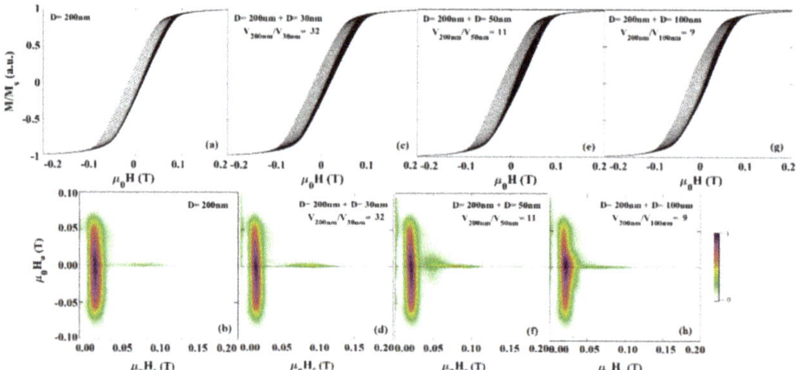

Figure 16. First-order reversal curve (FORC) data (**a,c,e,g**) and heat-maps (**b,d,f,h**) of magnetic nanowires in polycarbonate templates [143]. The horizontal and vertical axes of the heat-maps are the coercivity axes and interaction field axes, respectively.

4.2.13. AC Field Technique

The AC field technique measures the hysteresis loops of magnetic nanoparticles at high frequencies. The AC field technique was first proposed to characterize the applied field rate effects on the coercivity, a critical parameter for proper designing magnetic recording media. This technique became a very important characterization for understanding the heating efficiency of the magnetic nanoparticles, especially the superparamagnetic nanoparticles because they have zero-coercivity at static field that increases with the frequency of the applied field. The coercivity of the superparamagnetic nanoparticles is linearly proportional to the frequency of the applied fields for low frequencies while it behaves nonlinearly at high frequencies. This frequency dependence of the coercivity of the superparamagnetic nanoparticles was found as a great source of heat generation for biomedical applications, such as hyperthermia [58,150,151] or nanowarming [152]. For example, Nemati et al. reported on the shape and size effects on the heat generation of magnetic nanoparticles using the AC field where it was shown the magnetic nanodiscs have relatively larger heating efficiency compared to the nanospherical nanoparticles [70]. Elsewhere, Das et al. reported on the enhanced heating efficiency of elongated magnetic nanoparticles using the AC field [139], where they showed the shape anisotropy induced by elongation is the key for tuning the heating efficiency of the magnetic nanoparticles. The requirement of a high heating generation rate for special applications, such as nanowarming of cryopreserved organs, the AC field technique became a strong technique for characterizing the magnetic nanowires, as they are capable to generate much higher heats granted to their shape anisotropy [137,153].

5. Applications

5.1. Drug Delivery

In traditional drug delivery approaches, either intravascular injection or oral ingestion, the blood circulation distributes the medication throughout the body. This drug delivery mechanism causes only a small portion of the medicine to reach the targeted tissue or organ while the majority of the drug is distributed to healthy sites. Due to the small therapeutic windows for effectiveness of a drug, this mechanism of drug delivery is not efficient as it requires injection of high doses to assure the effectiveness of the drug, which can cause side-effects on healthy tissues/organs. In this particular application, the magnetic nanoparticles are highly desired as they can be controlled and directed to the targeted sites using an external field [154] (Figure 17). The movement of the magnetic nanoparticles can be controlled using either a static magnetic field or an alternating magnetic field. Considering the costs and maneuvering flexibility, the static field is significantly cheaper but less accurate compared to the alternating field approach. Another advantage of magnetic nanoparticles in drug delivery is their capability for generating heat. In order to control the drug dissociation or activation, the loaded drugs on the magnetic nanoparticles will not be released unless temperature increases sufficiently to activate it [155]. As an example, Kennedy et al. reported in magnetic regulation of drug delivery using the ferrogels [156]. The authors showed that the stimulation of biphasic ferrogels using an optimized magnetic signal can control the drug release rates of multiple drugs at different frequencies.

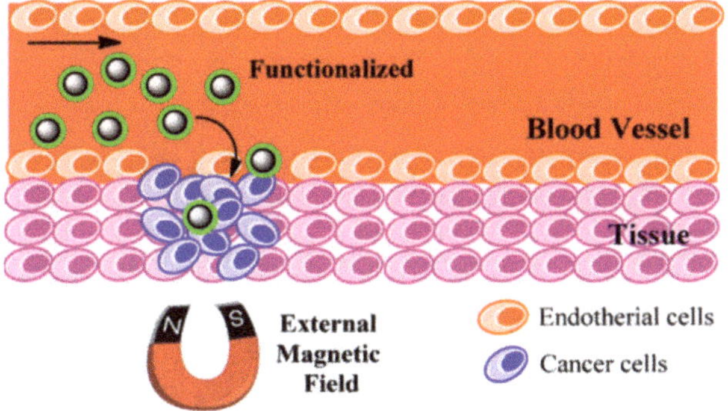

Figure 17. A schematic of magnetic nanoparticles for drug delivery. The magnetic nanoparticles concentrate at the targeted site using an external highly uniform and strong static magnetic field [30].

Magnetic nanoparticles must have high magnetization and anisotropy to induce sufficiently large electromotive power to overcome the drag forces imposed by the blood. In this prospect, the elongated magnetic nanoparticles, especially the magnetic nanowires, have superiority over the other magnetic nanoparticles, such as nanospheres [157]. The magnetic nanowires with large magnetic moment and anisotropy were shown to penetrate deeply into the tissues targeting favorable sites. For example, Pondman and co-authors reported on non-toxic drug delivery using the magnetic nanowires that were surface functionalized in oleic acid [155]. In another study, Alsharif et al. proposed using the surface functionalized core-shell iron-iron oxide nanowire using BSA for drug delivery [158]. The larger magnetic moment of the core-shell iron-iron oxide improved the controlling the magnetic nanowires; however, due to the larger magnetic interactions, these magnetic nanowires show larger degrees of aggregation leading to reducing the drug delivery efficiency.

5.2. Cancer Therapy: Thermal or Mechanical

One of the most interesting applications of magnetic nanoparticles is cancer therapy [159]. Magnetic nanoparticles have shown great promise in destroying cancerous cells via both thermal damage and mechanical damage. In thermal damage, also known as hyperthermia, three mechanisms are used to destroy the cancer cells: (1) Brownian mechanism, (2) Neel mechanism, and (3) hysteresis loss (Figure 18).

The Brownian mechanism damages the cancer cells by heat generation as a result of mechanical rotation and/or vibration, also known as external friction. The Neel mechanism damages cancer cells using the heat generation induced by the movement of the magnetic domains, which can be a result of the electromagnetic wave stimulus. The hysteresis loss, on the other hand, damages cancer cells using the heat generation induced by the switching the magnetic moment, as a result of an external magnetic field. Comparing the Neel mechanism and hysteresis loss, the hysteresis loss has significant advantage over the Neel mechanism as it can generate much higher heat rate. Comparing the Brownian mechanism and hysteresis loss, the hysteresis loss is practically more favorable because of its high selectivity for damaging the cancer cells with minimal damage to healthy ones compared to the Brownian mechanism [160].

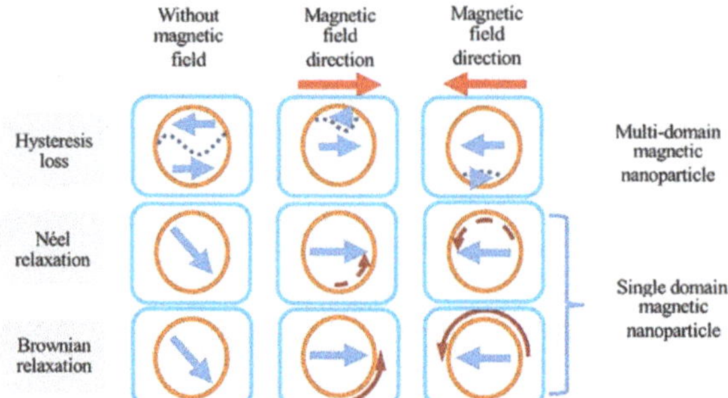

Figure 18. A schematic of the different heat generation modes using magnetic nanoparticles in cancer hyperthermia [161]. Orange circles represent magnetic nanoparticles, the straight arrows render magnetic field direction, curved arrows show the movement or change in magnetization direction (dashed curved arrow), and dashed lines illustrate domain boundaries in magnetic nanoparticles.

In this particular application, the anisotropy and saturation magnetization of the nanoparticles play a critical role on their efficiency [150]. As a result, it was shown that the nanocubes, nanoflakes, nano-octahedra, elongated and nanodisc nanoparticles have significantly higher hyperthermia efficiency compared to nanospheres [162,163]. For example, Goiriena-Goikoetxea et al., reported on high yield synthesis of permalloy nanodiscs with zero remanence and high magnetic moment for cancer therapy [164], where the authors used the hole-mask colloidal lithography method to cost-effectively and uniformly produce nanodiscs. Larger hysteresis loss can be achieved by the nanoparticles with larger hysteresis area, such as electrodeposited ferromagnetic nanowires. For hysteresis loss, the external magnetic field must be sufficiently larger to meet the coercivity of the nanoparticles [77]. Since ferromagnetic nanowires have very large coercivities, they require a very large external field to generate heat. Superparamagnetic nanoparticles, on the other hand, have zero coercivity at static fields while it increases as the frequency of the external magnetic field increases [29]. Thus, even though the ferromagnetic nanowires have a significant higher heating efficiency, the superparamagnetic nanoparticles provide more flexibility for designing a more effective hyperthermia cancer therapy because the magnetic field and frequency cannot exceed the biological safety limits, as they can cause severe damage to the healthy cells/tissues.

5.3. Cryopreservation

Cryopreservation is another form of heat generation using magnetic nanoparticles; however, in contrast to the hyperthermia, it involves heat generation from cryogenic temperatures up to the melting temperature of the cryopreservation agent. In contrast to hyperthermia, the only heat generation mechanism is the hysteresis loss because the magnetic nanoparticles are not able to produce any mechanical rotation or vibration as they are fixed at cryogenic temperatures. The main challenge in this application is generating enough heat over a short time leading to a high heat generation rate. If the heat generation rate is not sufficiently large to meet the critical warming rate of the cryopreservation agent, large thermal stresses will form across the tissue/organ. These thermal stresses induce large the mechanical stresses that can cause mechanical cracks, and ultimately destroying the cryopreserved tissue/organ [165,166]. Spherical superparamagnetic nanoparticles have shown promising results as they meet the critical warming rate of VS55, a commonly used cryopreservation agent. Since VS55 is toxic, low concentrations are highly desirable to minimize its side-effects after transplanting. The lower concertation of the VS55 requires a higher heat generation rate that is very challenging to be met. In this case, other magnetic nanoparticles, especially ferromagnetic nanowires, are superb candidates [137] as they are able to generate heat at significantly higher rates. General speaking, since the ferromagnetic nanowires have large coercivities, they require large external magnetic field which makes their implementation problematic because at large magnetic fields the eddy currents cause extremely non-uniform heat distributions. This particular application of magnetic nanoparticles has been recently proposed and still much efforts are needed to elucidate the most effective nanoparticles, especially in the case of large cryopreserved tissues/organs that required very uniform temperature increase.

5.4. MRI Imaging

MRI images are captured using a strong magnetic field, magnetic field gradient, and radio waves. The main advantage of the MRI imaging is that this technique does not require X-rays or ionizing radiation compared to CT scans and PET scans. However, it is very challenging to take MRI images from tissues/organs with similar magnetic properties [165]. In these cases, magnetic nanoparticles have been used as contrast agents to distinguish those tissues/organs by manipulating the relaxivities, transverse (T2) and/or longitudinal (T1), of their molecules. The T1 and T2 relaxation times indicate the path that molecules/tissues return to their equilibrium state, transverse to the static magnetic field and parallel to the static magnetic field, respectively. The magnetic nanoparticles were found to be a promising candidate, especially for manipulating the T2 relaxation time that was found to be more effective compared to manipulating the T1 relaxation time, due to their non-zero magnetic moment. The most favorable T2 relaxation time can be achieved using the magnetic nanoparticles with highest local magnetic inhomogeneities. Thus, elongated nanoparticles, such as nanorods and nanowires, are good candidates for this application as they have higher magnetic inhomogeneity because of their high surface to volume ratio [167]. For example, Smolensky et al. reported a trend of increased T1 relaxation time with increasing the diameter of the nanospheres [168]. In another example, Zhou and co-authors reported on magnetic nanodiscs with enhanced T1 relaxation times compared to magnetic nanospheres [169]. In this direction, Shore et al. introduced Fe/Au multi-segmented magnetic nanowires where they observed the shorter segments with larger diameters have significant improved T2 relaxation time [167]. More recently, Martinez-Banderas et al. introduced core-shell iron-iron oxide magnetic nanowires with excellent T2 relaxation time suitable for cell tracking [170]. The authors varied the oxidization level of their electrodeposited Fe nanowires and they showed the oxide thickness can be adjusted to tailor the transverse relaxation time.

5.5. Tissue Engineering

Tissue engineering is one of the major biomedical applications that has benefitted drastically from the capability of magnetic nanoparticles for being remotely actuated and/or controlled [171,172]. It was shown that engineering the texture of the tissues can successively improve nerve growth or cause cancer cell proliferation and migration to cease [173]. Traditionally, the fibrils of collagen-based tissues were oriented by applying a large magnetic field because the collagen is diamagnetic and it reorients perpendicular to the applied field direction [174]. Due to the requirement of very large magnetic field, in order of several thousand Oersted, this method is limited to a very small volume of the tissue. To overcome this limitation, Sharma et al. proposed crosslinking the collagen fibrils with surface functionalized magnetic nanowires prior to gelation. Using this technique, they showed that one-step bi-directional alignment of the collagen fibrils using only a few hundred Oersted [175] (Figure 19).

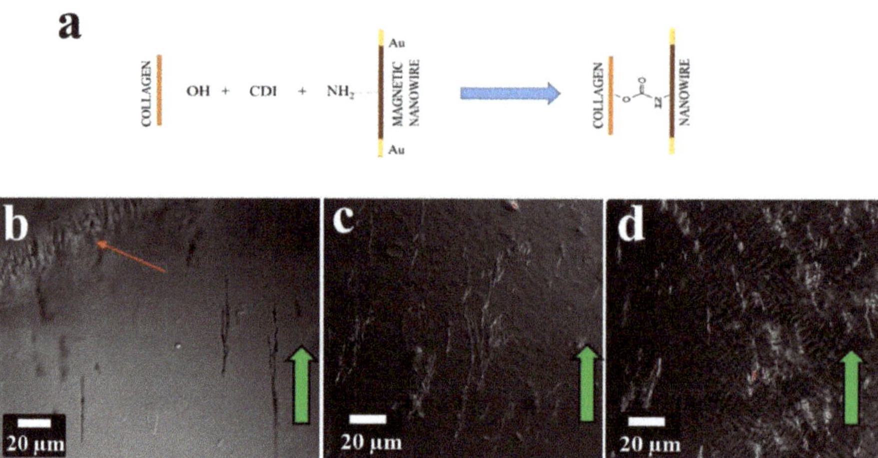

Figure 19. (**a**) A schematic of magnetic nanoparticles' application for tissue engineering via crosslinking magnetic nanowires with collagen fibrils under a highly uniform magnetic field during gelation [176]; (**b**) a diffraction contrast image, (**c**) dark-field image, and (**d**) confocal reflectance image.

Magnetic nanoparticles are also incorporated into hydrogels for producing ferrogels by chemically crosslinking the magnetic nanoparticles to the hydrogels. Ferrogel-based magnetic composites have attracted special attention, particularly in biosensing applications. An interesting work on the synthesis and comprehensive characterization of ferrogels was reported by Blyakhman et al. [177]. The authors synthesized and fully characterized two different series of ferrogels based on polyacrylamides with different chemical network densities. In another example, Safronov et al. reported on the synthesis of magnetically enriched polyacrylamide ferrogels by free radical polymerization using large aspect ratio magnetic nanowires, where they showed that large volume homogeneous magnetically enriched ferrogel-based tissues are feasible [5].

In addition to taking advantage of the remote actuation characteristics of magnetic nanoparticles for engineering regenerative tissues, magnetic nanoparticles also have been used in this particular application for inducing cells and delivering drugs promoting neuroregeneration. For example, Yuan et al. recently developed Au-coated superparamagnetic iron oxide nanoparticles to stimulate the nerve growth factor with low toxicity [176]. They showed that the coated magnetic nanoparticles provided higher neuronal growth and controlled orientation if they are stimulated under a dynamic magnetic field compared to a static magnetic field. Another interesting study was done by Polakka-Kanthikeel et al., where they showed that the surface functionalized magnetic nanoparticles are able to pass through the blood brain barrier to promote brain tissue repair [178]. Further information regarding the application and impacts of the magnetic nanoparticles in tissue engineering can be found in [173].

5.6. Enrichment

Magnetic nanoparticles are promising enrichment agents compared to optical nanoparticles [2,179]. This is because the magnetically enriched biological entities, such as cells, can be rapidly and cost-effectively separated and isolated from the whole population using a simple magnet [160] (Figure 20).

This can be readily done using a simple magnetic stand as opposed to costly, slow, and high complex optical systems. Magnetic nanospheres, nanocubes, and nano-octahedra have been widely used for this application because they can be mass produced cheaply. Furthermore, their rounded geometry, compared to the elongated nanoparticles, promotes their circulation inside capillaries with lower risk of blocking. However, there are two main challenges with these nanoparticles. First, they have relatively very small magnetic moment compared to ferromagnetic nanowires that makes it difficult to fully collect them. Second, ferromagnetic nanowires have shown a higher internalization yield that makes them superior for effective collect specific RNA/DNA and exosomes from targeted cells. For example, Hultgren et al. showed that nickel magnetic nanowires have an internalization yield of 35% larger than those of iron oxide nanospheres [180,181]. In another example, Sharma et al. observed high internalization of the gold-tipped nickel nanowires into osteosarcoma cells showing high cell activities and proliferation even after internalization of the magnetic nanowires [182,183]. In this direction, Nemati et al. enriched canine osteosarcoma cells (OSCA 8, 32, and 40) using iron-gold multi-segmented nanowires and they successfully collected the cancer-derived exosomes with a yield competitive to the commercialized methods [4].

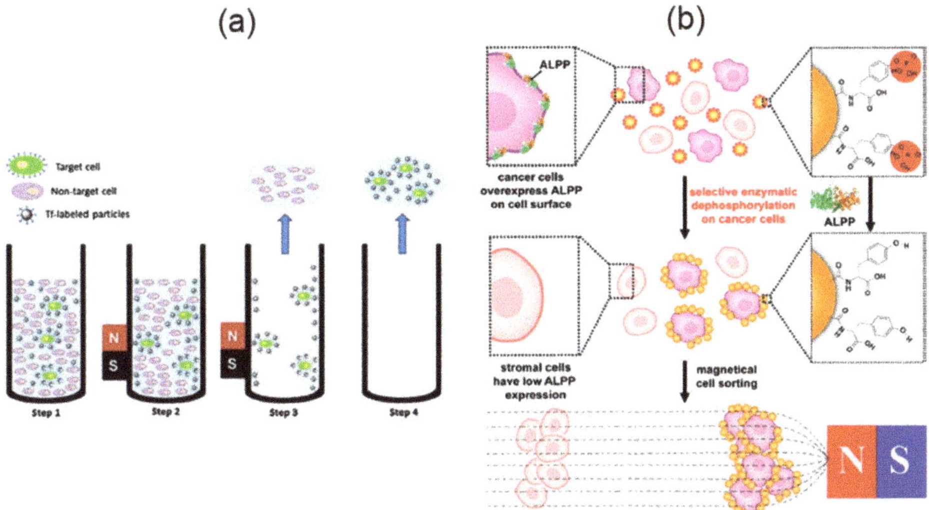

Figure 20. A schematic of the application of magnetic nanoparticles in enrichment [184]. (**a**) Scheme of cell separation using magnetic particles. Target cells are bound to magnetic particles modified with transferrin as a targeting moiety (step 1). Cells are magnetically separated (step 2). Non-targeted cells are removed with supernatant (step 3). Subsequently, target cells are re-suspended and removed (step 4). (**b**) Enzymatic transformation of magnetic particles for selective sorting of cancer cells.

5.7. Multiplexing/Demultiplexing Biolabels

Multiplexing/demultiplexing is crucial for many diseases, and tumors in particular, as they are often inhomogeneous. Multiplexing/demultiplexing can ensure both successful diagnosis and phenotyping of the discovered tumor [143]. Numerous attempts have been made to conjugate fluorophore molecules with magnetic nanoparticles to provide multiplexing. Since the multiplexing/demultiplexing of magnetic nanoparticles relies on the optical molecules, they inherently suffer the limitations of optical molecules [185]. The magnetic anisotropy of magnetic nanoparticles is a key for this application because it produces different magnetic signatures in different directions. Elongated magnetic nanoparticles, and in particular electrodeposited nanowires, exhibit a superior advantage over all other magnetic nanoparticles. That is because the electrodeposition technique provides much more flexibility to tailor the magnetic anisotropy via electrodepositing multi-segmented, multi-component, and multi-diameter magnetic nanowires with high-aspect ratios. For example, Sharma at al. electrodeposited multi-segmented magnetic nanowires for barcoding the cancer cells [183,184,186]. They used the hysteresis loop for detecting the barcodes, which suffers from low specificity [187]. Hysteresis loops only measure the saturation magnetization and the coercivity, which does not provide flexibility to generate multitude of unique detection.

As an alternative, Wen et al. designed a ferromagnetic resonance system to detect the resonance frequency of the magnetic nanowires as biolabels [188]. The two main limitations of this approach are: (1) the magnetic nanowires must be in contact with a co-planar waveguide, a condition which cannot be met in practical applications, and (2) the magnetic nanowires must be aligned as their ferromagnetic signature changes in different directions. More recently, Zamani Kouhpanji et al. proposed an approach that suppresses the former limitations, as it can detect the magnetic nanowires regardless of its location in the biological tissue [143]. Using FORC measurements and proposing a novel approach for demultiplexing the overlap magnetic signatures, they were able to successfully quantify the magnetic nanowires with high precision competitive to optical techniques. In another significant technological development, Zamani Kouhpanji et al. introduced a novel approach to significantly speed up the measurement while collecting more accurate data than FORC technique to more precisely quantify the magnetic nanowires in complex combinations [54] (Figure 21).

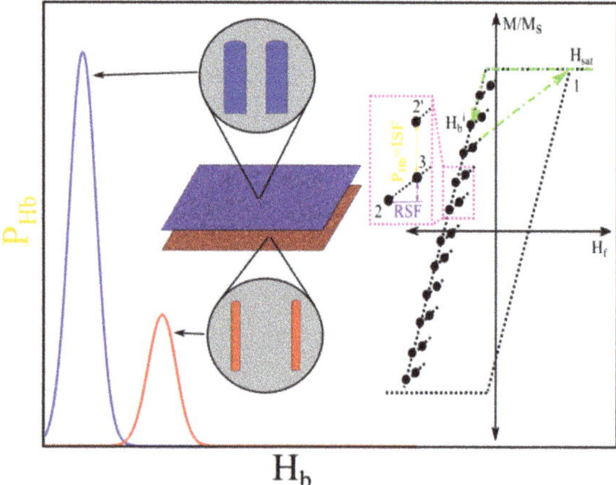

Figure 21. A schematic of the use of magnetic nanoparticles for multiplexing/demultiplexing [54]. The magnetic nanowires with different magnetic signatures, granted by their size and composition, were synthesized in polycarbonate tissues. The projection method, which directly measures the irreversible magnetization, was used to demultiplex the magnetically enriched tissues in complex combinations.

5.8. Biosensing and Biosensors

Magnetic nanoparticles have broadly used in diverse nanosensors, especially biosensors [189]. The surface functionalization of the magnetic nanoparticles, such as aptamer-modified magnetic nanoparticles, is a critical key in biosensing as it not only enhances their biocompatibility but also their specificity [190]. The magnetic interaction between the magnetic nanoparticles is the basis of their application in biosensors granted to their large surface to volume ratio leading to very high binding affinity [191]. These interaction forces substantially decrease the time response of these nanosensors. In the majority of biosensors, the magnetic nanoparticles are suspended in a solution and there are surface functionalized to detect the target molecules (Figure 22). An early report of using the magnetic nanoparticles for sensing was by Wang at al. who utilized magnetic nanoparticles to detect DNA with high selectivity and sensitivity [192]. More recent studies, for example Tavallaie et al., reported on sensing the RNA with extremely low concentration, as low as 1×10^{-17}M, using magnetic nanoparticles [193].

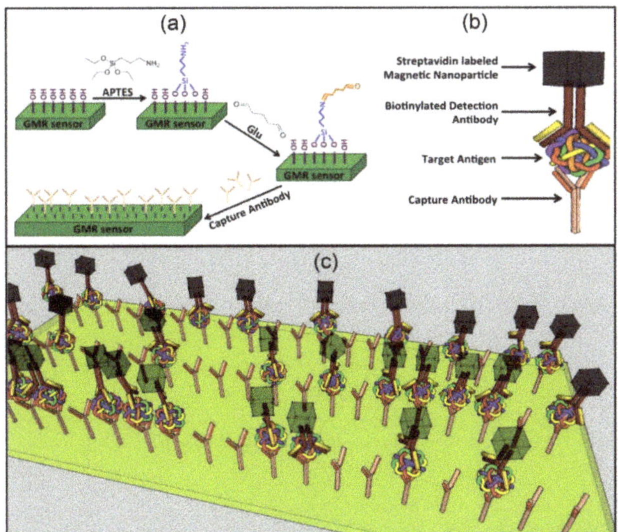

Figure 22. A schematic of the use of magnetic nanoparticles for biosensing applications [194]. In subfigures, (**a**) is a schematic diagram of a surface functionalized giant magnetoresistance (GMR) biosensor, (**b**) is the coated magnetic nanoparticles with a secondary antibody to detect the primary antibody on the surface of the GMR biosensor shown in (**c**).

For this particular application, Maruyama et al. reported on a fully automated biosensors using biomagnetic nanoparticles [195]. The authors were able to detect and identify two types of somatic mutations of epidermal growth factor receptor gene in lung cancer in a limited time. Their results are interesting as they indicate a very low mutation rate could be detected using magnetic nanoparticles. The giant magnetoimpedance biosensors and giant magnetoresistance biosensors have been widely used for detection in biomedical applications [196]. A pioneering work on giant magnetoresistance biosensors was reported by Baselt et al. [197]. In this work, the authors were able to measure intermolecular force binding, such as DNA-DNA, ligand-receptor, and antibody-antigen, using their portable giant magnetoresistance biosensors. Blyakhman et al. developed a giant magnetoimpedance sensor prototype for studying ferrogels [176]. These forms of biosensors are very promising for dispersed magnetic nanoparticles in solutions or ferrogels. However, detecting the magnetic nanoparticles incorporated into living entities, such as cells or biological tissues, is extremely difficult and it still not properly understood.

6. Summary/Outlook

In this review, we have provided a short, but still comprehensive review, of the synthesis and characterization techniques of magnetic nanoparticles. The main focus of this review was to understand the most effective approach for synthesizing and characterizing magnetic nanoparticles suitable for specific bio-applications, particularly biosensing and nanomedicine. The sensitivity of a particular bio-application on the monodispersity or polydispersity of the magnetic property is the key to choosing the most efficient route for synthesis and characterizing the required magnetic nanoparticles. For example, in biosensing applications, the magnetic nanoparticles do not necessarily need to have an identical morphology. However, the final achievable magnetic moment and size, which determine the magnetic nanoparticles binding affinity, and costs for mass-production are the keys determining the synthesis technique. On the other hand, for a bio-application such as multiplexing where the identical magnetic property is in high demand, the synthesis technique must be capable of precisely tuning the magnetic properties. As another example, for drug delivery and tissue engineering applications it is critical to have magnetic nanoparticles with a strong magnetic moment to assure their controllability in addition to a large surface area to load a sufficient amount of drugs. In this particular application, the elongated magnetic nanoparticles, especially electrodeposited magnetic nanowires, are of great interest as they generally have a strong magnetic moment with a larger surface area. The same situation is valid for selecting the most appropriate characterization technique. For example, in hyperthermia cancer therapy and cryopreservation applications, the heating efficiency of the magnetic nanoparticles is the key that must be adequately characterized and verified using AC field, zero-field cooling/field cooling (ZFC/FC), and magnetic susceptibility measurements. Particularly for biosensing applications, the characterization techniques such as Fourier transform infrared spectroscopy (FTIR) or thermal gravimetric analysis (TGA) are very useful as they provide valuable information regarding the amount and types of the coatings. To date, even though the magnetic nanoparticles have been proposed as promising multimodal nanoparticles in diverse nanobiotechnology applications, there are still numerous broken connections among those that must be addressed to realize the multimodality of the magnetic nanoparticles.

Author Contributions: Conceptualization, M.R.Z.K. and B.J.H.S.; Writing-Original Draft Preparation, M.R.Z.K.; Writing-Review & Editing, M.R.Z.K. and B.J.H.S.; Supervision, B.J.H.S.; Funding Acquisition, B.J.H.S. All authors have read and agreed to the published version of the manuscript.

Funding: This research was funded by National Science Foundation (NSF), grant number CMMI-1762884.

Conflicts of Interest: The authors declare no conflict of interest.

References

1. Salati, A.; Ramazani, A.; Almasi Kashi, M. Tuning hyperthermia properties of FeNiCo ternary alloy nanoparticles by morphological and magnetic characteristics. *J. Magn. Magn. Mater.* **2020**, *498*, 166172. [CrossRef]

2. Shore, D.E.; Dileepan, T.; Modiano, J.F.; Jenkins, M.K.; Stadler, B.J.H. Enrichment and Quantification of Epitope-Specific CD4+ T Lymphocytes Using Ferromagnetic Iron-Gold and Nickel Nanowires. *Sci. Rep.* **2018**, *8*, 15696. [CrossRef]

3. Alonso, J.; Khurshid, H.; Sankar, V.; Nemati, Z.; Phan, M.H.; Garayo, E.; García, J.A.; Srikanth, H. FeCo Nanowires with Enhanced Heating Powers and Controllable Dimensions for Magnetic Hyperthermia. *J. Appl. Phys.* **2015**, *117*, 17D113. [CrossRef]

4. Nemati, Z.; Um, J.; Zamani Kouhpanji, M.R.; Zhou, F.; Gage, T.; Shore, D.; Makielski, K.; Donnelly, A.; Alonso, J. Magnetic Isolation of Cancer-Derived Exosomes Using Fe/Au Magnetic Nanowires. *ACS Appl. Nano Mater.* **2020**, *3*, 2058–2069. [CrossRef]

5. Safronov, A.P.; Stadler, B.J.H.; Um, J.; Zamani Kouhpanji, M.R.; Alonso Masa, J.; Galyas, A.G.; Kurlyandskaya, G.V. Polyacrylamide Ferrogels with Ni Nanowires. *Materials* **2019**, *12*, 2582. [CrossRef] [PubMed]

6. Cui, H.; Shi, J.; Yuan, B.; Fu, M. Synthesis of Porous Magnetic Ferrite Nanowires Containing Mn and Their Application in Water Treatment. *J. Mater. Chem. A* **2013**, *1*, 5902–5907. [CrossRef]

7. Chio, D.S.; Park, J.; Kim, S.; Gracias, D.H.; Cho, M.K.; Kim, Y.K.; Fung, A.; Lee, S.E.; Chen, Y.; Khanal, S.; et al. Hyperthermia with Magnetic Nanowires for Inactivating Living Cells. *J. Nanosci. Nanotechnol.* **2008**, *8*, 2323–2327. [CrossRef]

8. Huang, X.; Tan, L.; Cho, H.; Stadler, B.J.H. Magnetoresistance and spin transfer torque in electrodeposited Co/Cu multilayered nanowire arrays with small diameters. *J. Appl. Phys.* **2009**, *105*, 1–4. [CrossRef]

9. Maqableh, M.M.; Huang, X.; Sung, S.; Reddy, K.S.M.; Norby, G.; Victora, R.H.; Stadler, B.J.H. Low-Resistivity 10 Nm Diameter Magnetic Sensors. *Nano Lett.* **2012**, *12*, 4102–4109. [CrossRef]

10. Parkin, S.; Yang, S.-H. Memory on the Racetrack. *Nat. Nanotechnol.* **2015**, *10*, 195–198. [CrossRef]

11. Parkin, S.S.P.; Hayashi, M.; Thomas, L. Magnetic Domain-Wall Racetrack Memory. *Science* **2008**, *320*, 190–195. [CrossRef] [PubMed]

12. Qiu, Y.; Myers, D.R.; Lam, W.A. The Biophysics and Mechanics of Blood from a Materials Perspective. *Nat. Rev. Mater.* **2019**, *4*, 294–311. [CrossRef]

13. Zhu, H.; Zhang, L.; Tong, S.; Lee, C.M.; Deshmukh, H.; Bao, G. Spatial Control of in Vivo CRISPR-Cas9 Genome Editing via Nanomagnets. *Nat. Biomed. Eng.* **2019**, *3*, 126–136. [CrossRef] [PubMed]

14. Peng, F.; Setyawati, M.I.; Tee, J.K.; Ding, X.; Wang, J.; Nga, M.E.; Ho, H.K.; Leong, D.T. Nanoparticles Promote in Vivo Breast Cancer Cell Intravasation and Extravasation by Inducing Endothelial Leakiness. *Nat. Nanotechnol.* **2019**, *14*, 279–286. [CrossRef]

15. Lau, Y.; Betto, D.; Rode, K.; Coey, J.M.D.; Stamenov, P. Spin-Orbit Torque Switching without an External Field Using Interlayer Exchange Coupling. *Nat. Nanotechnol.* **2016**, *11*, 758–762. [CrossRef] [PubMed]

16. Tsymbal, E.Y.; Kohlstedt, H. Tunneling Across a Ferroelectric. *Science* **2006**, *313*, 181–184. [CrossRef]

17. Baumgartner, M.; Garello, K.; Mendil, J.; Avci, C.O.; Grimaldi, E.; Murer, C.; Feng, J.; Gabureac, M.; Stamm, C.; Acremann, Y.; et al. Spatially and Time-Resolved Magnetization Dynamics Driven by Spin-Orbit Torques. *Nat. Nanotechnol.* **2017**, *12*, 980–986. [CrossRef]

18. Koyama, T.; Chiba, D.; Ueda, K.; Kondou, K.; Tanigawa, H.; Fukami, S.; Suzuki, T.; Ohshima, N.; Ishiwata, N.; Nakatani, Y.; et al. Observation of the Intrinsic Pinning of a Magnetic Domain Wall in a Ferromagnetic Nanowire. *Nat. Mater.* **2011**, *10*, 194–197. [CrossRef]

19. Brataas, A.; Kent, A.D.; Ohno, H. Current-Induced Torques in Magnetic Materials. *Nat. Mater.* **2012**, *11*, 372–381. [CrossRef]

20. Bauer, U.; Emori, S.; Beach, G.S.D. Voltage-Controlled Domain Wall Traps in Ferromagnetic Nanowires. *Nat. Nanotechnol.* **2013**, *8*, 411–416. [CrossRef]

21. Dadfar, S.M.; Roemhild, K.; Drude, N.I.; von Stillfried, S.; Knüchel, R.; Kiessling, F.; Lammers, T. Iron Oxide Nanoparticles: Diagnostic, Therapeutic and Theranostic Applications. *Adv. Drug Deliv. Rev.* **2019**, *138*, 302–325. [CrossRef] [PubMed]

22. Sajid, M.; Płotka-Wasylka, J. Nanoparticles: Synthesis, Characteristics, and Applications in Analytical and Other Sciences. *Microchem. J.* **2020**, 104623. [CrossRef]

23. Wu, S.; Sun, A.; Zhai, F.; Wang, J.; Xu, W.; Zhang, Q.; Volinsky, A.A. Fe_3O_4 Magnetic Nanoparticles Synthesis from Tailings by Ultrasonic Chemical Co-Precipitation. *Mater. Lett.* **2011**, *65*, 1882–1884. [CrossRef]

24. Pereira, C.; Pereira, A.M.; Fernandes, C.; Rocha, M.; Mendes, R.; Fernández-García, M.P.; Guedes, A.; Tavares, P.B.; Grenèche, J.-M.; Araújo, J.P.; et al. Superparamagnetic MFe_2O_4 (M = Fe, Co, Mn) Nanoparticles: Tuning the Particle Size and Magnetic Properties through a Novel One-Step Coprecipitation Route. *Chem. Mater.* **2012**, *24*, 1496–1504. [CrossRef]

25. Petcharoen, K.; Sirivat, A. Synthesis and Characterization of Magnetite Nanoparticles via the Chemical Co-Precipitation Method. *Mater. Sci. Eng. B* **2012**, *177*, 421–427. [CrossRef]

26. Alonso, J.; Barandiarán, J.M.; Fernández Barquín, L.; García-Arribas, A. Magnetic Nanoparticles, Synthesis, Properties, and Applications. In *Magnetic Nanostructured Materials*; Elsevier: Amsterdam, The Netherlands, 2018; pp. 1–40. [CrossRef]

27. Unni, M.; Uhl, A.M.; Savliwala, S.; Savitzky, B.H.; Dhavalikar, R.; Garraud, N.; Arnold, D.P.; Kourkoutis, L.F.; Andrew, J.S.; Rinaldi, C. Thermal Decomposition Synthesis of Iron Oxide Nanoparticles with Diminished Magnetic Dead Layer by Controlled Addition of Oxygen. *ACS Nano* **2017**, *11*, 2284–2303. [CrossRef]

28. Lee, N.; Hyeon, T. Designed Synthesis of Uniformly Sized Iron Oxide Nanoparticles for Efficient Magnetic Resonance Imaging Contrast Agents. *Chem. Soc. Rev.* **2012**, *41*, 2575–2589. [CrossRef]

29. Nemati, Z.; Alonso, J.; Rodrigo, I.; Das, R.; Garaio, E.; García, J.Á.; Orue, I.; Phan, M.H.; Srikanth, H. Improving the Heating Efficiency of Iron Oxide Nanoparticles by Tuning Their Shape and Size. *J. Phys. Chem. C* **2018**, *122*, 2367–2381. [CrossRef]

30. Wu, W.; Wu, Z.; Yu, T.; Jiang, C.; Kim, W.-S. Recent Progress on Magnetic Iron Oxide Nanoparticles: Synthesis, Surface Functional Strategies and Biomedical Applications. *Sci. Technol. Adv. Mater.* **2015**, *16*, 023501. [CrossRef]

31. Laudise, R.A. *Hydrothermal Synthesis of Crystals 50 Years Progress in Crystal Growth*; Elsevier: Amsterdam, The Netherlands, 2004. [CrossRef]

32. Chen, Y.; Xia, H.; Lu, L.; Xue, J. Synthesis of Porous Hollow Fe_3O_4 Beads and Their Applications in Lithium Ion Batteries. *J. Mater. Chem.* **2012**, *22*, 5006. [CrossRef]

33. Torres-Gómez, N.; Nava, O.; Argueta-Figueroa, L.; García-Contreras, R.; Baeza-Barrera, A.; Vilchis-Nestor, A.R. Shape Tuning of Magnetite Nanoparticles Obtained by Hydrothermal Synthesis: Effect of Temperature. *J. Nanomater.* **2019**, *2019*, 1–15. [CrossRef]

34. Dippong, T.; Cadar, O.; Levei, E.A.; Bibicu, I.; Diamandescu, L.; Leostean, C.; Lazar, M.; Borodi, G.; Tudoran, L.B. Structure and Magnetic Properties of $CoFe_2O_4/SiO_2$ Nanocomposites Obtained by Sol-Gel and Post Annealing Pathways. *Ceram. Int.* **2017**, *43*, 2113–2122. [CrossRef]

35. Laurent, S.; Forge, D.; Port, M.; Roch, A.; Robic, C.; Vander Elst, L.; Muller, R.N. Magnetic Iron Oxide Nanoparticles: Synthesis, Stabilization, Vectorization, Physicochemical Characterizations, and Biological Applications. *Chem. Rev.* **2008**, *108*, 2064–2110. [CrossRef] [PubMed]

36. Darbandi, M.; Stromberg, F.; Landers, J.; Reckers, N.; Sanyal, B.; Keune, W.; Wende, H. Nanoscale Size Effect on Surface Spin Canting in Iron Oxide Nanoparticles Synthesized by the Microemulsion Method. *J. Phys. D Appl. Phys.* **2012**, *45*, 195001. [CrossRef]

37. Praça, F.G.; Viegas, J.S.R.; Peh, H.Y.; Garbin, T.N.; Medina, W.S.G.; Bentley, M.V.L.B. Microemulsion Co-Delivering Vitamin A and Vitamin E as a New Platform for Topical Treatment of Acute Skin Inflammation. *Mater. Sci. Eng. C* **2020**, *110*, 110639. [CrossRef] [PubMed]

38. Argüelles-Pesqueira, A.I.; Diéguez-Armenta, N.M.; Bobadilla-Valencia, A.K.; Nataraj, S.K.; Rosas-Durazo, A.; Esquivel, R.; Alvarez-Ramos, M.E.; Escudero, R.; Guerrero-German, P.; Lucero-Acuña, J.A.; et al. Low Intensity Sonosynthesis of Iron Carbide@iron Oxide Core-Shell Nanoparticles. *Ultrason. Sonochem.* **2018**, *49*, 303–309. [CrossRef]

39. Yadav, R.S.; Kuřitka, I.; Vilcakova, J.; Jamatia, T.; Machovsky, M.; Skoda, D.; Urbánek, P.; Masař, M.; Urbánek, M.; Kalina, L.; et al. Impact of Sonochemical Synthesis Condition on the Structural and Physical Properties of $MnFe_2O_4$ Spinel Ferrite Nanoparticles. *Ultrason. Sonochem.* **2020**, *61*, 104839. [CrossRef]

40. Thickett, S.C.; Teo, G.H. Recent Advances in Colloidal Nanocomposite Design via Heterogeneous Polymerization Techniques. *Polym. Chem.* **2019**, *10*, 2906–2924. [CrossRef]

41. Majidi, S.; Sehrig, F.Z.; Farkhani, S.M.; Soleymani Goloujeh, M.; Akbarzadeh, A. Current Methods for Synthesis of Magnetic Nanoparticles. *Artif. Cells Nanomed. Biotechnol.* **2016**, *44*, 722–734. [CrossRef]

42. Teo, B.M.; Chen, F.; Hatton, T.A.; Grieser, F.; Ashokkumar, M. Novel One-Pot Synthesis of Magnetite Latex Nanoparticles by Ultrasound Irradiation. *Langmuir* **2009**, *25*, 2593–2595. [CrossRef] [PubMed]

43. Hu, X.; Yu, J.C.; Gong, J.; Li, Q.; Li, G. α-Fe_2O_3 Nanorings Prepared by a Microwave-Assisted Hydrothermal Process and Their Sensing Properties. *Adv. Mater.* **2007**, *19*, 2324–2329. [CrossRef]

44. Kezuka, Y.; Yoshida, M.; Tajika, M. Template-Free Fabrication of Single-Crystalline Calcite Nanorings during Crystal Growth in Water. *Cryst. Eng. Comm.* **2020**, *22*, 9–13. [CrossRef]

45. Liu, S.; Lu, F.; Jia, X.; Cheng, F.; Jiang, L.-P.; Zhu, J.-J. Microwave-Assisted Synthesis of a Biocompatible Polyacid-Conjugated Fe_3O_4 Superparamagnetic Hybrid. *CrystEngComm* **2011**, *13*, 2425. [CrossRef]

46. Aivazoglou, E.; Metaxa, E.; Hristoforou, E. Microwave-Assisted Synthesis of Iron Oxide Nanoparticles in Biocompatible Organic Environment. *AIP Adv.* **2018**, *8*, 048201. [CrossRef]

47. Qiu, G.; Huang, H.; Genuino, H.; Opembe, N.; Stafford, L.; Dharmarathna, S.; Suib, S.L. Microwave-Assisted Hydrothermal Synthesis of Nanosized α-Fe_2O_3 for Catalysts and Adsorbents. *J. Phys. Chem. C* **2011**, *115*, 19626–19631. [CrossRef]

48. Piraux, L. Magnetic Nanowires. *Appl. Sci.* **2020**, *10*, 1832. [CrossRef]

49. Reddy, S.M.; Park, J.J.; Na, S.-M.; Maqableh, M.M.; Flatau, A.B.; Bethanie, J.H. Electrochemical Synthesis of Magnetostrictive Fe-Ga/Cu Multilayered Nanowire Arrays with Tailored Magnetic Response. *Adv. Funct. Mater.* **2011**, *21*, 4677–4683. [CrossRef]

50. Araujo, E.; Encinas, A.; Velázquez-Galván, Y.; Martínez-Huerta, J.M.; Hamoir, G.; Ferain, E.; Piraux, L. Artificially Modified Magnetic Anisotropy in Interconnected Nanowire Networks. *Nanoscale* **2015**, *7*, 1485–1490. [CrossRef]

51. Lee, W.; Ji, R.; Ross, C.A.; Gösele, U.; Nielsch, K. Wafer-Scale Ni Imprint Stamps for Porous Alumina Membranes Based on Interference Lithography. *Small* **2006**, *2*, 978–982. [CrossRef]

52. Um, J.; Zamani Kouhpanji, M.R.; Liu, S.; Nemati, Z.; Kosel, J.; Stadler, B. Fabrication of Long-Range Ordered Aluminum Oxide and Fe/Au Multilayered Nanowires for 3D Magnetic Memory. *IEEE Trans. Magn.* **2020**, *56*, 1–6. [CrossRef]

53. Sulka, G.D.; Brzózka, A.; Liu, L. Fabrication of Diameter-Modulated and Ultrathin Porous Nanowires in Anodic Aluminum Oxide Templates. *Electrochim. Acta* **2011**, *56*, 4972–4979. [CrossRef]

54. Kouhpanji, M.R.Z.; Stadler, B.J.H. Projection Method as a Probe for Multiplexing/Demultiplexing of Magnetically Enriched Biological Tissues. *RSC Adv.* **2020**, *10*, 13286–13292. [CrossRef]

55. Cabrera, L.; Gutierrez, S.; Menendez, N.; Morales, M.P.; Herrasti, P. Magnetite Nanoparticles: Electrochemical Synthesis and Characterization. *Electrochim. Acta* **2008**, *53*, 3436–3441. [CrossRef]

56. Rodríguez-López, A.; Paredes-Arroyo, A.; Mojica-Gomez, J.; Estrada-Arteaga, C.; Cruz-Rivera, J.J.; Elías Alfaro, C.G.; Antaño-López, R. Electrochemical Synthesis of Magnetite and Maghemite Nanoparticles Using Dissymmetric Potential Pulses. *J. Nanoparticle Res.* **2012**, *14*, 993. [CrossRef]

57. Li, X.; Xu, H.; Chen, Z.-S.; Chen, G. Biosynthesis of Nanoparticles by Microorganisms and Their Applications. *J. Nanomater.* **2011**, *2011*, 1–16. [CrossRef]

58. Marcano, L.; Muñoz, D.; Martín-Rodríguez, R.; Orue, I.; Alonso, J.; García-Prieto, A.; Serrano, A.; Valencia, S.; Abrudan, R.; Fernández Barquín, L.; et al. Magnetic Study of Co-Doped Magnetosome Chains. *J. Phys. Chem. C* **2018**, *122*, 7541–7550. [CrossRef]

59. Walle, A.V.D.; Sangnier, A.P.; Abou-Hassan, A.; Curcio, A.; Hémadi, M.; Menguy, N.; Lalatonne, Y.; Luciani, N.; Wilhelm, C. Biosynthesis of Magnetic Nanoparticles from Nano-Degradation Products Revealed in Human Stem Cells. *Proc. Natl. Acad. Sci. USA* **2019**, *116*, 4044–4053. [CrossRef]

60. Bharde, A.A.; Parikh, R.Y.; Baidakova, M.; Jouen, S.; Hannoyer, B.; Enoki, T.; Prasad, B.L.V.; Shouche, Y.S.; Ogale, S.; Sastry, M. Bacteria-Mediated Precursor-Dependent Biosynthesis of Superparamagnetic Iron Oxide and Iron Sulfide Nanoparticles. *Langmuir* **2008**, *24*, 5787–5794. [CrossRef]

61. Bazylinski, D.A.; Frankel, R.B. Magnetosome Formation in Prokaryotes. *Nat. Rev. Microbiol.* **2004**, *2*, 217–230. [CrossRef]

62. Kotov, Y.A. Electric Explosion of Wires as a Method for Preparation of Nanopowders. *J. Nanoparticle Res.* **2003**, *5*, 539–550. [CrossRef]

63. Safronov, A.P.; Beketov, I.V.; Komogortsev, S.V.; Kurlyandskaya, G.V.; Medvedev, A.I.; Leiman, D.V.; Larrañaga, A.; Bhagat, S.M. Spherical Magnetic Nanoparticles Fabricated by Laser Target Evaporation. *AIP Adv.* **2013**, *3*, 052135. [CrossRef]

64. Vergés, M.A.; Costo, R.; Roca, A.G.; Marco, J.F.; Goya, G.F.; Serna, C.J.; Morales, M.P. Uniform and Water Stable Magnetite Nanoparticles with Diameters around the Monodomain–Multidomain Limit. *J. Phys. D Appl. Phys.* **2008**, *41*, 134003. [CrossRef]

65. Roca, A.G.; Gutiérrez, L.; Gavilán, H.; Brollo, M.E.F.; Veintemillas-Verdaguer, S.; Morales, M.D.P. Design Strategies for Shape-Controlled Magnetic Iron Oxide Nanoparticles. *Adv. Drug Deliv. Rev.* **2019**, *138*, 68–104. [CrossRef]

66. Kovalenko, M.V.; Bodnarchuk, M.I.; Lechner, R.T.; Hesser, G.; Schäffler, F.; Heiss, W. Fatty Acid Salts as Stabilizers in Size- and Shape-Controlled Nanocrystal Synthesis: The Case of Inverse Spinel Iron Oxide. *J. Am. Chem. Soc.* **2007**, *129*, 6352–6353. [CrossRef] [PubMed]

67. Yang, H.; Ogawa, T.; Hasegawa, D.; Takahashi, M. Synthesis and Magnetic Properties of Monodisperse Magnetite Nanocubes. *J. Appl. Phys.* **2008**, *103*, 07D526. [CrossRef]

68. Gavilán, H.; Posth, O.; Bogart, L.K.; Steinhoff, U.; Gutiérrez, L.; Morales, M.P. How Shape and Internal Structure Affect the Magnetic Properties of Anisometric Magnetite Nanoparticles. *Acta Mater.* **2017**, *125*, 416–424. [CrossRef]

69. Yang, Y.; Liu, X.; Lv, Y.; Herng, T.S.; Xu, X.; Xia, W.; Zhang, T.; Fang, J.; Xiao, W.; Ding, J. Orientation Mediated Enhancement on Magnetic Hyperthermia of Fe_3O_4 Nanodisc. *Adv. Funct. Mater.* **2015**, *25*, 812–820. [CrossRef]

70. Nemati, Z.; Salili, S.M.; Alonso, J.; Ataie, A.; Das, R.; Phan, M.H.; Srikanth, H. Superparamagnetic Iron Oxide Nanodiscs for Hyperthermia Therapy: Does Size Matter? *J. Alloys Compd.* **2017**, *714*, 709–714. [CrossRef]

71. Chen, L.; Yang, X.; Chen, J.; Liu, J.; Wu, H.; Zhan, H.; Liang, C.; Wu, M. Continuous Shape- and Spectroscopy-Tuning of Hematite Nanocrystals. *Inorg. Chem.* **2010**, *49*, 8411–8420. [CrossRef]

72. Palchoudhury, S.; An, W.; Xu, Y.; Qin, Y.; Zhang, Z.; Chopra, N.; Holler, R.A.; Turner, C.H.; Bao, Y. Synthesis and Growth Mechanism of Iron Oxide Nanowhiskers. *Nano Lett.* **2011**, *11*, 1141–1146. [CrossRef]

73. Geng, S.; Yang, H.; Ren, X.; Liu, Y.; He, S.; Zhou, J.; Su, N.; Li, Y.; Xu, C.; Zhang, X.; et al. Anisotropic Magnetite Nanorods for Enhanced Magnetic Hyperthermia. *Chem. Asian J.* **2016**, *11*, 2996–3000. [CrossRef] [PubMed]

74. Rebolledo, A.F.; Bomatí-Miguel, O.; Marco, J.F.; Tartaj, P. A Facile Synthetic Route for the Preparation of Superparamagnetic Iron Oxide Nanorods and Nanorices with Tunable Surface Functionality. *Adv. Mater.* **2008**, *20*, 1760–1765. [CrossRef]

75. Wiogo, H.; Lim, M.; Munroe, P.; Amal, R. Understanding the Formation of Iron Oxide Nanoparticles with Acicular Structure from Iron(III) Chloride and Hydrazine Monohydrate. *Cryst. Growth Des.* **2011**, *11*, 1689–1696. [CrossRef]

76. Ozaki, M.; Kratohvil, S.; Matijević, E. Formation of Monodispersed Spindle-Type Hematite Particles. *J. Colloid Interface Sci.* **1984**, *102*, 146–151. [CrossRef]

77. Nana, A.B.A.; Marimuthu, T.; Kondiah, P.P.D.; Choonara, Y.E.; Toit, L.C.D.; Pillay, V. Multifunctional Magnetic Nanowires: Design, Fabrication, and Future Prospects as Cancer Therapeutics. *Cancers* **2019**, *11*, 1956. [CrossRef]

78. Tang, B.; Wang, G.; Zhuo, L.; Ge, J.; Cui, L. Facile Route to α-FeOOH and α-Fe$_2$O$_3$ Nanorods and Magnetic Property of α-Fe$_2$O$_3$ Nanorods. *Inorg. Chem.* **2006**, *45*, 5196–5200. [CrossRef]

79. Kim, B.H.; Yang, J.; Lee, D.; Choi, B.K.; Hyeon, T.; Park, J. Liquid-Phase Transmission Electron Microscopy for Studying Colloidal Inorganic Nanoparticles. *Adv. Mater.* **2018**, *30*, 1703316. [CrossRef]

80. Sandler, S.E.; Fellows, B.; Mefford, O.T. Best Practices for Characterization of Magnetic Nanoparticles for Biomedical Applications. *Anal. Chem.* **2019**, *91*, 14159–14169. [CrossRef]

81. Mourdikoudis, S.; Pallares, R.M.; Thanh, N.T.K. Characterization Techniques for Nanoparticles: Comparison and Complementarity upon Studying Nanoparticle Properties. *Nanoscale* **2018**, *10*, 12871–12934. [CrossRef]

82. Liao, H.-G.; Cui, L.; Whitelam, S.; Zheng, H. Real-Time Imaging of Pt3Fe Nanorod Growth in Solution. *Science* **2012**, *336*, 1011–1014. [CrossRef]

83. Kato, H.; Suzuki, M.; Fujita, K.; Horie, M.; Endoh, S.; Yoshida, Y.; Iwahashi, H.; Takahashi, K.; Nakamura, A.; Kinugasa, S. Reliable Size Determination of Nanoparticles Using Dynamic Light Scattering Method for in Vitro Toxicology Assessment. *Toxicol. Vitr.* **2009**, *23*, 927–934. [CrossRef] [PubMed]

84. Lim, J.; Yeap, S.P.; Che, H.X.; Low, S.C. Characterization of Magnetic Nanoparticle by Dynamic Light Scattering. *Nanoscale Res. Lett.* **2013**, *8*, 381. [CrossRef] [PubMed]

85. Hole, P.; Sillence, K.; Hannell, C.; Maguire, C.M.; Roesslein, M.; Suarez, G.; Capracotta, S.; Magdolenova, Z.; Horev-Azaria, L.; Dybowska, A.; et al. Interlaboratory Comparison of Size Measurements on Nanoparticles Using Nanoparticle Tracking Analysis (NTA). *J. Nanoparticle Res.* **2013**, *15*, 2101. [CrossRef] [PubMed]

86. Zhao, X.; Zhang, W.; Qiu, X.; Mei, Q.; Luo, Y.; Fu, W. Rapid and Sensitive Exosome Detection with CRISPR/Cas12a. *Anal. Bioanal. Chem.* **2020**, *412*, 601–609. [CrossRef] [PubMed]

87. Ribeiro, L.N.; de Couto, V.M.; Fraceto, L.F.; de Paula, E. Use of Nanoparticle Concentration as a Tool to Understand the Structural Properties of Colloids. *Sci. Rep.* **2018**, *8*, 982. [CrossRef]

88. Upadhyay, S.; Parekh, K.; Pandey, B. Influence of Crystallite Size on the Magnetic Properties of Fe$_3$O$_4$ Nanoparticles. *J. Alloys Compd.* **2016**, *678*, 478–485. [CrossRef]

89. Li, W.; Zamani, R.; Rivera Gil, P.; Pelaz, B.; Ibáñez, M.; Cadavid, D.; Shavel, A.; Alvarez-Puebla, R.A.; Parak, W.J.; Arbiol, J.; et al. CuTe Nanocrystals: Shape and Size Control, Plasmonic Properties, and Use as SERS Probes and Photothermal Agents. *J. Am. Chem. Soc.* **2013**, *135*, 7098–7101. [CrossRef]

90. Sharma, S.K. Complex Magnetic Nanostructures. Sharma, S.K., Ed.; Springer: Cham, Switzerland, 2017. [CrossRef]

91. Blanco, A.C. Sodium Carbonate Mediated Synthesis of Iron Oxide Nanoparticles to Improve Magnetic Hyperthermia Efficiency and Induce Apoptosis. Ph.D. Thesis, University College London, London, UK, January 2014; p. 600.

92. Shukla, N.; Liu, C.; Jones, P.M.; Weller, D. FTIR Study of Surfactant Bonding to FePt Nanoparticles. *J. Magn. Magn. Mater.* **2003**, *266*, 178–184. [CrossRef]

93. Tzitzios, V.; Basina, G.; Gjoka, M.; Alexandrakis, V.; Georgakilas, V.; Niarchos, D.; Boukos, N.; Petridis, D. Chemical Synthesis and Characterization of Hcp Ni Nanoparticles. *Nanotechnology* **2006**, *17*, 3750–3755. [CrossRef]

94. Lu, L.T. Water-Dispersible Magnetic Nanoparticles for Biomedical Applications: Synthesis and Characterisation. Ph.D. Thesis, University of Liverpool, Liverpool, UK, January 2011.

95. Nan, A.; Suciu, M.; Ardelean, I.; Şenilă, M.; Turcu, R. Characterization of the Nuclear Magnetic Resonance Relaxivity of Gadolinium Functionalized Magnetic Nanoparticles. *Anal. Lett.* **2020**, 1–16. [CrossRef]

96. Mompeán, M.; Sánchez-Donoso, R.M.; de la Hoz, A.; Saggiomo, V.; Velders, A.H.; Gomez, M.V. Pushing Nuclear Magnetic Resonance Sensitivity Limits with Microfluidics and Photo-Chemically Induced Dynamic Nuclear Polarization. *Nat. Commun.* **2018**, *9*, 108. [CrossRef]

97. Silva Elipe, M.V. Advantages and Disadvantages of Nuclear Magnetic Resonance Spectroscopy as a Hyphenated Technique. *Anal. Chim. Acta* **2003**, *497*, 1–25. [CrossRef]

98. Andjelković, L.; Jeremić, D.; Milenković, M.R.; Radosavljević, J.; Vulić, P.; Pavlović, V.; Manojlović, D.; Nikolić, A.S. Synthesis, Characterization and in Vitro Evaluation of Divalent Ion Release from Stable $NiFe_2O_4$, $ZnFe_2O_4$ and Core-Shell $ZnFe_2O_4@NiFe_2O_4$ Nanoparticles. *Ceram. Int.* **2020**, *46*, 3528–3533. [CrossRef]
99. Harkness, K.M.; Cliffel, D.E.; McLean, J.A. Characterization of Thiolate-Protected Gold Nanoparticles by Mass Spectrometry. *Analyst* **2010**, *135*, 868. [CrossRef] [PubMed]
100. Mortazavi-Manesh, A.; Bagherzadeh, M. Synthesis and Characterization of Molybdenum (VI) Complex Immobilized on Polymeric Schiff Base-coated Magnetic Nanoparticles as an Efficient and Retrievable Nanocatalyst in Olefin Epoxidation Reactions. *Appl. Organomet. Chem.* **2020**, 34. [CrossRef]
101. Pourjavadi, A.; Kohestanian, M.; Streb, C. PH and Thermal Dual-Responsive Poly(NIPAM-Co-GMA)-Coated Magnetic Nanoparticles via Surface-Initiated RAFT Polymerization for Controlled Drug Delivery. *Mater. Sci. Eng. C* **2020**, *108*, 110418. [CrossRef] [PubMed]
102. Karimi-Chayjani, R.; Daneshvar, N.; Nikoo Langarudi, M.S.; Shirini, F.; Tajik, H. Silica-Coated Magnetic Nanoparticles Containing Bis Dicationic Bridge for the Synthesis of 1,2,4-Triazolo Pyrimidine/Quinazolinone Derivatives. *J. Mol. Struct.* **2020**, *1199*, 126891. [CrossRef]
103. Rudolph, M.; Erler, J.; Peuker, U.A. A TGA–FTIR Perspective of Fatty Acid Adsorbed on Magnetite Nanoparticles–Decomposition Steps and Magnetite Reduction. Colloids Surfaces a Physicochem. *Eng. Asp.* **2012**, *397*, 16–23. [CrossRef]
104. Ziegler-Borowska, M.; Chełminiak, D.; Kaczmarek, H. Thermal Stability of Magnetic Nanoparticles Coated by Blends of Modified Chitosan and Poly(Quaternary Ammonium) Salt. *J. Therm. Anal. Calorim.* **2015**, *119*, 499–506. [CrossRef]
105. Cullity, B.D.; Graham, C.D. *Introduction to Magnetic Materials*; John Wiley & Sons: Hoboken, NJ, USA, 2009.
106. Kouhpanji, M.R.Z.; Stadler, B.J.H. Quantitative Description of Complex Magnetic Nanoparticle Arrays. *arXiv* **2019**, arXiv:1911.12480.
107. Kouhpanji, M.R.Z.; Stadler, B.J.H. Assessing the Reliability and Validity Ranges of Magnetic Characterization Methods. *arXiv* **2020**, arXiv:2003.06911.
108. Kouhpanji, M.R.Z.; Visscher, P.B.; Stadler, B.J.H. Underlying Magnetization Responses of Magnetic Nanoparticles in Assemblies. *arXiv* **2020**, arXiv:2002.07742.
109. Tadic, M.; Milosevic, I.; Kralj, S.; Hanzel, D.; Barudzija, T.; Motte, L.; Makovec, D. Surface-Induced Reversal of a Phase Transformation for the Synthesis of ε-Fe_2O_3 Nanoparticles with High Coercivity. *Acta Mater.* **2020**, *188*, 16–22. [CrossRef]
110. Bououdina, M.S.; Manoharan, C. Dependence of Structure/Morphology on Electrical/Magnetic Properties of Hydrothermally Synthesised Cobalt Ferrite Nanoparticles. *J. Magn. Magn. Mater.* **2020**, *493*, 165703. [CrossRef]
111. Oh, S.-J.; Choi, C.-J.; Kwon, S.-J.; Jin, S.-H.; Kim, B.-K.; Park, J.-S. Mössbauer Analysis on the Magnetic Properties of Fe–Co Nanoparticles Synthesized by Chemical Vapor Condensation Process. *J. Magn. Magn. Mater.* **2004**, *280*, 147–157. [CrossRef]
112. Bystrzejewski, M.; Grabias, A.; Borysiuk, J.; Huczko, A.; Lange, H. Mössbauer Spectroscopy Studies of Carbon-Encapsulated Magnetic Nanoparticles Obtained by Different Routes. *J. Appl. Phys.* **2008**, *104*, 054307. [CrossRef]
113. Tiano, A.L.; Papaefthymiou, G.C.; Lewis, C.S.; Han, J.; Zhang, C.; Li, Q.; Shi, C.; Abeykoon, A.M.M.; Billinge, S.J.L.; Stach, E.; et al. Correlating Size and Composition-Dependent Effects with Magnetic, Mössbauer, and Pair Distribution Function Measurements in a Family of Catalytically Active Ferrite Nanoparticles. *Chem. Mater.* **2015**, *27*, 3572–3592. [CrossRef]
114. Xiao, J.; Kuc, A.; Pokhrel, S.; Mädler, L.; Pöttgen, R.; Winter, F.; Frauenheim, T.; Heine, T. Fe-Doped ZnO Nanoparticles: The Oxidation Number and Local Charge on Iron, Studied by 57 Fe Mößbauer Spectroscopy and DFT Calculations. *Chem. A Eur. J.* **2013**, *19*, 3287–3291. [CrossRef]
115. Diehl, M.R.; Yu, J.-Y.; Heath, J.R.; Held, G.A.; Doyle, H.; Sun, S.; Murray, C.B. Crystalline, Shape, and Surface Anisotropy in Two Crystal Morphologies of Superparamagnetic Cobalt Nanoparticles by Ferromagnetic Resonance. *J. Phys. Chem. B* **2001**, *105*, 7913–7919. [CrossRef]
116. Herrling, M.P.; Fetsch, K.L.; Delay, M.; Blauert, F.; Wagner, M.; Franzreb, M.; Horn, H.; Lackner, S. Low Biosorption of PVA Coated Engineered Magnetic Nanoparticles in Granular Sludge Assessed by Magnetic Susceptibility. *Sci. Total Environ.* **2015**, *537*, 43–50. [CrossRef]

117. Kitamoto, Y.; He, J.-S. Chemical Synthesis of FePt Nanoparticles with High Alternate Current Magnetic Susceptibility for Biomedical Applications. *Electrochim. Acta* **2009**, *54*, 5969–5972. [CrossRef]

118. Ortega, E.; Reddy, S.M.; Betancourt, I.; Roughani, S.; Stadler, B.J.H.; Ponce, A. Magnetic Ordering in 45 Nm-Diameter Multisegmented FeGa/Cu Nanowires: Single Nanowires and Arrays. *J. Mater. Chem. C* **2017**, *5*, 7546–7552. [CrossRef]

119. Biziere, N.; Gatel, C.; Clochard, M.C.; Wegrowe, J.E.; Snoeck, E. Imaging the Fine Structure of a Magnetic Domain Wall in a Ni Nanocylinder. *Nano Lett.* **2013**, *13*, 2053–2057. [CrossRef]

120. Cantu-Valle, J.; Betancourt, I.; Sanchez, J.E.; Ruiz-Zepeda, F.; Maqableh, M.M.; Mendoza-Santoyo, F.; Stadler, B.J.H.; Ponce, A. Mapping the Magnetic and Crystal Structure in Cobalt Nanowires. *J. Appl. Phys.* **2015**, *118*, 024302. [CrossRef] [PubMed]

121. Robertson, D.J.; France, D.E. Discrimination of Remanence-Carrying Minerals in Mixtures, Using Isothermal Remanent Magnetisation Acquisition Curves. *Phys. Earth Planet. Int.* **1994**, *82*, 223–234. [CrossRef]

122. Kelly, P.E.; O'Grady, K.; Mayo, P.L.; Chantrell, R.W. Switching Mechanisms in Cobalt-Phosphorus Thin Films. *IEEE Trans. Magn.* **1989**, *25*, 3881–3883. [CrossRef]

123. Huerta, J.M.M.; De La Torre Medina, J.; Piraux, L.; Encinas, A. Self Consistent Measurement and Removal of the Dipolar Interaction Field in Magnetic Particle Assemblies and the Determination of Their Intrinsic Switching Field Distribution. *J. Appl. Phys.* **2012**, *111*. [CrossRef]

124. Araujo, E.; Martínez-Huerta, J.M.; Piraux, L.; Encinas, A. Quantification of the Interaction Field in Arrays of Magnetic Nanowires from the Remanence Curves. *J. Supercond. Nov. Magn.* **2018**, *31*, 3981–3987. [CrossRef]

125. Moya, C.; Iglesias, Ó.; Batlle, X.; Labarta, A. Quantification of Dipolar Interactions in $Fe_{3-x}O_4$ Nanoparticles. *J. Phys. Chem. C* **2015**, *119*, 24142–24148. [CrossRef]

126. Del Bianco, L.; Spizzo, F.; Barucca, G.; Marangoni, G.; Sgarbossa, P. Glassy Magnetic Behavior and Correlation Length in Nanogranular Fe-Oxide and Au/Fe-Oxide Samples. *Materials* **2019**, *12*, 3958. [CrossRef]

127. De Toro, J.A.; Vasilakaki, M.; Lee, S.S.; Andersson, M.S.; Normile, P.S.; Yaacoub, N.; Murray, P.; Sánchez, E.H.; Muñiz, P.; Peddis, D.; et al. Remanence Plots as a Probe of Spin Disorder in Magnetic Nanoparticles. *Chem. Mater.* **2017**, *29*, 8258–8268. [CrossRef]

128. Che, X.; Neal Bertram, H. Phenomenology of ΔM Curves and Magnetic Interactions. *J. Magn. Magn. Mater.* **1992**, *116*, 121–127. [CrossRef]

129. Gräfe, J.; Schmidt, M.; Audehm, P.; Schütz, G.; Goering, E. Application of Magneto-Optical Kerr Effect to First-Order Reversal Curve Measurements. *Rev. Sci. Instrum.* **2014**, *85*. [CrossRef] [PubMed]

130. Palmero, E.M.; Bran, C.; del Real, R.P.; Vázquez, M. Vortex Domain Wall Propagation in Periodically Modulated Diameter FeCoCu Nanowire as Determined by the Magneto-Optical Kerr Effect. *Nanotechnology* **2015**, *26*, 461001. [CrossRef]

131. Bran, C.; Berganza, E.; Fernandez-Roldan, J.A.; Palmero, E.M.; Meier, J.; Calle, E.; Jaafar, M.; Foerster, M.; Aballe, L.; Fraile Rodriguez, A.; et al. Magnetization Ratchet in Cylindrical Nanowires. *ACS Nano* **2018**, *12*, 5932–5939. [CrossRef]

132. Burn, D.M.; Arac, E.; Atkinson, D. Magnetization Switching and Domain-Wall Propagation Behavior in Edge-Modulated Ferromagnetic Nanowire Structures. *Phys. Rev. B Condens. Matter Mater. Phys.* **2013**, *88*, 1–8. [CrossRef]

133. Mohammed, H.; Corte-León, H.; Ivanov, Y.P.; Lopatin, S.; Moreno, J.A.; Chuvilin, A.; Salimath, A.; Manchon, A.; Kazakova, O.; Kosel, J. Current Controlled Magnetization Switching in Cylindrical Nanowires for High-Density 3D Memory Applications. *arxiv* **2018**, arXiv:1804.06616.

134. Nasirpouri, F. Template Electrodeposition of Nanowires Arrays. In *Springer Series in Surface Sciences*; Springer: Berlin/Heidelberg, Germany, 2017; Volume 62, pp. 187–259. [CrossRef]

135. Passeri, D.; Dong, C.; Reggente, M.; Angeloni, L.; Barteri, M.; Scaramuzzo, F.A.; De Angelis, F.; Marinelli, F.; Antonelli, F.; Rinaldi, F.; et al. Magnetic Force Microscopy: Quantitative issues in biomaterials. *Biomatter* **2014**, *4*, e29507. [CrossRef]

136. Coïsson, M.; Barrera, G.; Celegato, F.; Manzin, A.; Vinai, F.; Tiberto, P. Magnetic Vortex Chirality Determination via Local Hysteresis Loops Measurements with Magnetic Force Microscopy. *Sci. Rep.* **2016**, *6*, 1–9. [CrossRef]

137. Shore, D.; Ghemes, A.; Dragos-Pinzaru, O.; Gao, Z.; Shao, Q.; Sharma, A.; Um, J.; Tabakovic, I.; Bischof, J.C.; Stadler, B.J.H. Nanowarming Using Au-Tipped $Co_{35}Fe_{65}$ Ferromagnetic Nanowires. *Nanoscale* **2019**, *11*, 14607–14615. [CrossRef]

138. Khusrhid, H.; Nemati Porshokouh, Z.; Phan, M.-H.; Mukherjee, P.; Srikanth, H. Impacts of Surface Spins and Inter-Particle Interactions on the Magnetism of Hollow γ-Fe$_2$O$_3$ Nanoparticles. *J. Appl. Phys.* **2014**, *115*, 17E131. [CrossRef]

139. Das, R.; Alonso, J.; Nemati Porshokouh, Z.; Kalappattil, V.; Torres, D.; Phan, M.H.; Garaio, E.; García, J.Á.; Llamazares, J.L.S.; Srikanth, H. Tunable High Aspect Ratio Iron Oxide Nanorods for Enhanced Hyperthermia. *J. Phys. Chem. C* **2016**, *120*, 10086–10093. [CrossRef]

140. Das, R.; Rinaldi-Montes, N.; Alonso, J.; Amghouz, Z.; Garaio, E.; García, J.A.; Gorria, P.; Blanco, J.A.; Phan, M.H.; Srikanth, H. Boosted Hyperthermia Therapy by Combined AC Magnetic and Photothermal Exposures in Ag/Fe$_3$O$_4$ Nanoflowers. *ACS Appl. Mater. Interfaces* **2016**, *8*, 25162–25169. [CrossRef] [PubMed]

141. Aharoni, A. Angular Dependence of Nucleation by Curling in a Prolate Spheroid. *J. Appl. Phys.* **2014**, *82*, 1281–1287. [CrossRef]

142. Chen, Y.; Xu, C.; Zhou, Y.; Maaz, K.; Yao, H.; Mo, D.; Lyu, S.; Duan, J.; Liu, J. Temperature- and Angle-Dependent Magnetic Properties of Ni Nanotube Arrays Fabricated by Electrodeposition in Polycarbonate Templates. *Nanomaterials* **2016**, *6*, 231. [CrossRef]

143. Kouhpanji, M.R.Z.; Um, J.; Stadler, B.J.H. Demultiplexing of Magnetic Nanowires with Overlapping Signatures for Tagged Biological Species. *ACS Appl. Nano Mater.* **2020**, *3*, 3080–3087. [CrossRef]

144. Kouhpanji, M.R.Z.; Stadler, B.J.H. Beyond the Qualitative Description of Complex Magnetic Nanoparticle Arrays Using FORC Measurement. *Nano Express* **2020**, *1*, 010017. [CrossRef]

145. Roberts, A.P.; Heslop, D.; Zhao, X.; Pike, C.R. Understanding Fine Magnetic Particle Systems through Use of First-Order Reversal Curve Diagrams. *Am. Geophys. Union* **2014**, *52*, 557–602. [CrossRef]

146. Ramazani, A.; Asgari, V.; Montazer, A.H.; Kashi, M.A. Tuning Magnetic Fingerprints of FeNi Nanowire Arrays by Varying Length and Diameter. *Curr. Appl. Phys.* **2015**, *15*, 819–828. [CrossRef]

147. Berndt, T.A.; Chang, L. Waiting for Forcot: Accelerating FORC Processing 100× Using a Fast-Fourier-Transform Algorithm. Geochemistry, Geophys. *Geosystems* **2019**, *20*, 6223–6233. [CrossRef]

148. Groß, F.; Martínez-García, J.C.; Ilse, S.E.; Schütz, G.; Goering, E.; Rivas, M.; Gräfe, J. GFORC: A Graphics Processing Unit Accelerated First-Order Reversal-Curve Calculator. *J. Appl. Phys.* **2019**, *126*, 163901. [CrossRef]

149. Cimpoesu, D.; Dumitru, I.; Stancu, A. DoFORC Tool for Calculating First-Order Reversal Curve Diagrams of Noisy Scattered Data. *J. Appl. Phys.* **2019**, 125. [CrossRef]

150. Alonso, J.; Khurshid, H.; Devkota, J.; Nemati, Z.; Khadka, N.K.; Srikanth, H.; Pan, J.; Phan, M.-H. Superparamagnetic Nanoparticles Encapsulated in Lipid Vesicles for Advanced Magnetic Hyperthermia and Biodetection. *J. Appl. Phys.* **2016**, *119*, 083904. [CrossRef]

151. Gandia, D.; Gandarias, L.; Rodrigo, I.; Robles-García, J.; Das, R.; Garaio, E.; García, J.Á.; Phan, M.; Srikanth, H.; Orue, I.; et al. Unlocking the Potential of Magnetotactic Bacteria as Magnetic Hyperthermia Agents. *Small* **2019**, *15*, 1902626. [CrossRef]

152. Manuchehrabadi, N.; Gao, Z.; Zhang, J.; Ring, H.L.; Shao, Q.; Liu, F.; Mcdermott, M.; Fok, A.; Rabin, Y.; Brockbank, K.G.M.; et al. Improved Tissue Cryopreservation Using Inductive Heating of Magnetic Nanoparticles. *Sci. Transl. Med.* **2017**, *9*, eaah4586. [CrossRef] [PubMed]

153. Marbaix, J.; Mille, N.; Lacroix, L.-M.; Asensio, J.M.; Fazzini, P.-F.; Soulantica, K.; Carrey, J.; Chaudret, B. Tuning the Composition of FeCo Nanoparticle Heating Agents for Magnetically Induced Catalysis. *ACS Appl. Nano Mater.* **2020**, *3*, 3767–3778. [CrossRef]

154. Xie, W.; Guo, Z.; Gao, F.; Gao, Q.; Wang, D.; Liaw, B.S.; Cai, Q.; Sun, X.; Wang, X.; Zhao, L. Shape-, Size- and Structure-Controlled Synthesis and Biocompatibility of Iron Oxide Nanoparticles for Magnetic Theranostics. *Theranostics* **2018**, *8*, 3284–3307. [CrossRef]

155. Pondman, K.M.; Bunt, N.D.; Maijenburg, A.W.; Van Wezel, R.J.A.; Kishore, U.; Abelmann, L.; Ten Elshof, J.E.; Ten Haken, B. Magnetic Drug Delivery with FePd Nanowires. *J. Magn. Magn. Mater.* **2015**, *380*, 299–306. [CrossRef]

156. Kennedy, S.; Roco, C.; Déléris, A.; Spoerri, P.; Cezar, C.; Weaver, J.; Vandenburgh, H.; Mooney, D. Improved Magnetic Regulation of Delivery Profiles from Ferrogels. *Biomaterials* **2018**, *161*, 179–189. [CrossRef]

157. Heidarshenas, B.; Wei, H.; Moghimi, Z.A.M.; Hussain, G.; Baniasadi, F.; Naghieh, G. Nanowires in Magnetic Drug Targeting. *Mater. Sci. Eng. Int. J.* **2019**, *3*, 3–9. [CrossRef]

158. Alsharif, N.A.; Martiinez-Banderas, A.; Merzaban, J.; Ravasi, T.; Kosel, J. Biofunctionalizing Magnetic Nanowires Toward Targeting and Killing Leukemia Cancer Cells. *IEEE Trans. Magn.* **2019**, *55*, 1–5. [CrossRef]

159. Grossman, J.H.; McNeil, S.E. Nanotechnology in Cancer Medicine. *Phys. Today* **2012**, *65*, 38–42. [CrossRef]
160. Serrà, A.; Vallés, E. Advanced Electrochemical Synthesis of Multicomponent Metallic Nanorods and Nanowires: Fundamentals and Applications. *Appl. Mater. Today* **2018**, *12*, 207–234. [CrossRef]
161. Chang, D.; Lim, M.; Goos, J.A.C.M.; Qiao, R.; Ng, Y.Y.; Mansfeld, F.M.; Jackson, M.; Davis, T.P.; Kavallaris, M. Biologically Targeted Magnetic Hyperthermia: Potential and Limitations. *Front. Pharmacol.* **2018**, 9. [CrossRef] [PubMed]
162. Mohapatra, J.; Mitra, A.; Aslam, M.; Bahadur, D. Octahedral-Shaped Fe3O4 Nanoparticles with Enhanced Specific Absorption Rate and R2 Relaxivity. *IEEE Trans. Magn.* **2015**, *51*, 19–21. [CrossRef]
163. Wang, F.; Li, C.; Cheng, J.; Yuan, Z. Recent Advances on Inorganic Nanoparticle-Based Cancer Therapeutic Agents. *Int. J. Environ. Res. Public Health* **2016**, *13*, 1182. [CrossRef] [PubMed]
164. Goiriena-Goikoetxea, M.; García-Arribas, A.; Rouco, M.; Svalov, A.V.; Barandiaran, J.M. High-Yield Fabrication of 60 Nm Permalloy Nanodiscs in Well-Defined Magnetic Vortex State for Biomedical Applications. *Nanotechnology* **2016**, *27*, 175302. [CrossRef]
165. Giwa, S.; Lewis, J.K.; Alvarez, L.; Langer, R.; Roth, A.E.; Church, G.M.; Markmann, J.F.; Sachs, D.H.; Chandraker, A.; Wertheim, J.A.; et al. The Promise of Organ and Tissue Preservation to Transform Medicine. *Nat. Biotechnol.* **2017**, *35*, 530–542. [CrossRef]
166. Bischof, J.C.; Diller, K.R. From Nanowarming to Thermoregulation: New Multiscale Applications of Bioheat Transfer. *Annu. Rev. Biomed. Eng.* **2018**, *20*, 301–327. [CrossRef]
167. Shore, D.; Pailloux, S.L.; Zhang, J.; Gage, T.; Flannigan, D.J.; Garwood, M.; Pierre, V.C.; Stadler, B.J.H. Electrodeposited Fe and Fe-Au Nanowires as MRI Contrast Agents. *Chem. Commun.* **2016**, *52*, 12634–12637. [CrossRef]
168. Smolensky, E.D.; Park, H.-Y.E.; Zhou, Y.; Rolla, G.A.; Marjańska, M.; Botta, M.; Pierre, V.C. Scaling Laws at the Nanosize: The Effect of Particle Size and Shape on the Magnetism and Relaxivity of Iron Oxide Nanoparticle Contrast Agents. *J. Mater. Chem. B* **2013**, *1*, 2818. [CrossRef] [PubMed]
169. Zhou, Z.; Bai, R.; Munasinghe, J.; Shen, Z.; Nie, L.; Chen, X. T1—T2 Dual-Modal Magnetic Resonance Imaging: From Molecular Basis to Contrast Agents. *ACS Nano* **2017**, *11*, 5227–5232. [CrossRef]
170. Martínez-Banderas, A.I.; Aires, A.; Plaza-García, S.; Colás, L.; Moreno, J.A.; Ravasi, T.; Merzaban, J.S.; Ramos-Cabrer, P.; Cortajarena, A.L.; Kosel, J. Magnetic Core–Shell Nanowires as MRI Contrast Agents for Cell Tracking. *J. Nanobiotechnol.* **2020**, *18*, 1–12. [CrossRef] [PubMed]
171. Ito, A.; Kamihira, M. *Tissue Engineering Using Magnetite Nanoparticles*, 1st ed.; Elsevier: Amsterdam, The Netherlands, 2011; Volume 104, pp. 355–395. [CrossRef]
172. Fathi-Achachelouei, M.; Knopf-Marques, H.; Ribeiro da Silva, C.E.; Barthès, J.; Bat, E.; Tezcaner, A.; Vrana, N.E. Use of Nanoparticles in Tissue Engineering and Regenerative Medicine. *Front. Bioeng. Biotechnol.* **2019**, *7*, 1–22. [CrossRef] [PubMed]
173. Ansari, S.A.M.K.; Ficiarà, E.; Ruffinatti, F.A.; Stura, I.; Argenziano, M.; Abollino, O.; Cavalli, R.; Guiot, C.; D'Agata, F. Magnetic Iron Oxide Nanoparticles: Synthesis, Characterization and Functionalization for Biomedical Applications in the Central Nervous System. *Materials* **2019**, *12*, 465. [CrossRef]
174. Noor Smal Löwik, D.W.P.M. Magnetic Fields to Align Natural and Synthetic Fibers. In *Self-Assembling Biomaterials*; Elsevier: Amsterdam, The Netherlands, 2018; pp. 321–340. [CrossRef]
175. Sharma, A.; DiVito, M.D.; Shore, D.E.; Block, A.D.; Pollock, K.; Solheid, P.; Feinberg, J.M.; Modiano, J.; Lam, C.H.; Hubel, A.; et al. Alignment of Collagen Matrices Using Magnetic Nanowires and Magnetic Barcode Readout Using First Order Reversal Curves (FORC) (Invited). *J. Magn. Magn. Mater.* **2018**, *459*, 176–181. [CrossRef]
176. Yuan, M.; Wang, Y.; Qin, Y.X. Promoting Neuroregeneration by Applying Dynamic Magnetic Fields to a Novel Nanomedicine: Superparamagnetic Iron Oxide (SPIO)-Gold Nanoparticles Bounded with Nerve Growth Factor (NGF). *Nanomed. Nanotechnol. Biol. Med.* **2018**, *14*, 1337–1347. [CrossRef]
177. Blyakhman, F.; Buznikov, N.; Sklyar, T.; Safronov, A.; Golubeva, E.; Svalov, A.; Sokolov, S.; Melnikov, G.; Orue, I.; Kurlyandskaya, G. Mechanical, Electrical and Magnetic Properties of Ferrogels with Embedded Iron Oxide Nanoparticles Obtained by Laser Target Evaporation: Focus on Multifunctional Biosensor Applications. *Sensors* **2018**, *18*, 872. [CrossRef]
178. Pilakka-Kanthikeel, S.; Atluri, V.S.R.; Sagar, V.; Saxena, S.K.; Nair, M. Targeted Brain Derived Neurotropic Factors (BDNF) Delivery across the Blood-Brain Barrier for Neuro-Protection Using Magnetic Nano Carriers: An In-Vitro Study. *PLoS ONE* **2013**, *8*, e62241. [CrossRef]

179. Jeon, Y.S.; Shin, H.M.; Kim, Y.J.; Nam, D.Y.; Park, B.C.; Yoo, E.; Kim, H.-R.; Kim, Y.K. Metallic Fe–Au Barcode Nanowires as a Simultaneous T Cell Capturing and Cytokine Sensing Platform for Immunoassay at the Single-Cell Level. *ACS Appl. Mater. Interfaces* **2019**, *11*, 23901–23908. [CrossRef]
180. Hultgren, A.; Tanase, M.; Chen, C.S.; Reich, D.H. High-Yield Cell Separations Using Magnetic Nanowires. *IEEE Trans. Magn.* **2004**, *40*, 2988–2990. [CrossRef]
181. Hultgren, A.; Tanase, M.; Chen, C.S.; Meyer, G.J.; Reich, D.H. Cell Manipulation Using Magnetic Nanowires. *J. Appl. Phys.* **2003**, *93*, 7554–7556. [CrossRef]
182. Sharma, A.; Orlowski, G.M.; Zhu, Y.; Shore, D.; Kim, S.Y.; DiVito, M.D.; Hubel, A.; Stadler, B.J.H. Inducing Cells to Disperse Nickel Nanowires via Integrin-Mediated Responses. *Nanotechnology* **2015**, *26*, 135102. [CrossRef]
183. Sharma, A. Multi-Segmented Magnetic Nanowires as Multifunctional Theranostic Tools in Nanomedicine. Ph.D. Thesis, University of Minnesota Twin Cities, Twin Cities, MN, USA, July 2015.
184. Kudr, J.; Haddad, Y.; Richtera, L.; Heger, Z.; Cernak, M.; Adam, V.; Zitka, O. Magnetic Nanoparticles: From Design and Synthesis to Real World Applications. *Nanomaterials* **2017**, *7*, 243. [CrossRef]
185. Bhana, S.; Wang, Y.; Huang, X. Nanotechnology for Enrichment and Detection of Circulating Tumor Cells. *Micro Nanosyst.* **2015**, *7*, 1973–1990. [CrossRef]
186. Sharma, A.; Zhu, Y.; Thor, S.; Zhou, F.; Stadler, B.; Hubel, A. Magnetic Barcode Nanowires for Osteosarcoma Cell Control, Detection and Separation. *IEEE Trans. Magn.* **2013**, *49*, 453–456. [CrossRef]
187. Shikha, S.; Salafi, T.; Cheng, J.; Zhang, Y. Versatile Design and Synthesis of Nano-Barcodes. *Chem. Soc. Rev.* **2017**, *46*, 7054–7093. [CrossRef]
188. Zhou, W.; Um, J.; Zhang, Y.; Nelson, A.P.; Nemati, Z.; Modiano, J.; Stadler, B.; Franklin, R. Development of a Biolabeling System Using Ferromagnetic Nanowires. *IEEE J. Electromagn. RF Microw. Med. Biol.* **2019**, *3*, 134–142. [CrossRef]
189. Megens, M.; Prins, M. Magnetic Biochips: A New Option for Sensitive Diagnostics. *J. Magn. Magn. Mater.* **2005**, *293*, 702–708. [CrossRef]
190. Modh, H.; Scheper, T.; Walter, J.-G. Aptamer-Modified Magnetic Beads in Biosensing. *Sensors* **2018**, *18*, 1041. [CrossRef]
191. Gloag, L.; Mehdipour, M.; Chen, D.; Tilley, R.D.; Gooding, J.J. Advances in the Application of Magnetic Nanoparticles for Sensing. *Adv. Mater.* **2019**, *31*, 1–26. [CrossRef] [PubMed]
192. Wang, J.; Xu, D.; Kawde, A.-N.; Polsky, R. Metal Nanoparticle-Based Electrochemical Stripping Potentiometric Detection of DNA Hybridization. *Anal. Chem.* **2001**, *73*, 5576–5581. [CrossRef] [PubMed]
193. Tavallaie, R.; McCarroll, J.; Le Grand, M.; Ariotti, N.; Schuhmann, W.; Bakker, E.; Tilley, R.D.; Hibbert, D.B.; Kavallaris, M.; Gooding, J.J. Nucleic Acid Hybridization on an Electrically Reconfigurable Network of Gold-Coated Magnetic Nanoparticles Enables MicroRNA Detection in Blood. *Nat. Nanotechnol.* **2018**, *13*, 1066–1071. [CrossRef]
194. Krishna, V.D.; Wu, K.; Perez, A.M.; Wang, J.-P. Giant Magnetoresistance-Based Biosensor for Detection of Influenza A Virus. *Front. Microbiol.* **2016**, *7*. [CrossRef]
195. Maruyama, K.; Takeyama, H.; Mori, T.; Ohshima, K.; Ogura, S.-I.; Mochizuki, T.; Matsunaga, T. Detection of Epidermal Growth Factor Receptor (EGFR) Mutations in Non-Small Cell Lung Cancer (NSCLC) Using a Fully Automated System with a Nano-Scale Engineered Biomagnetite. *Biosens. Bioelectron.* **2007**, *22*, 2282–2288. [CrossRef]
196. Blanc-Béguin, F.; Nabily, S.; Gieraltowski, J.; Turzo, A.; Querellou, S.; Salaun, P.Y. Cytotoxicity and GMI Bio-Sensor Detection of Maghemite Nanoparticles Internalized into Cells. *J. Magn. Magn. Mater.* **2009**, *321*, 192–197. [CrossRef]
197. Baselt, D.R.; Lee, G.U.; Natesan, M.; Metzger, S.W.; Sheehan, P.E.; Colton, R.J. A Biosensor Based on Magnetoresistance Technology. *Biosens. Bioelectron.* **1998**, *13*, 731–739. [CrossRef]

Article

Magnetoimpedance in Symmetric and Non-Symmetric Nanostructured Multilayers: A Theoretical Study

Nikita A. Buznikov [1] and Galina V. Kurlyandskaya [2,3,*]

[1] Scientific and Research Institute of Natural Gases and Gas Technologies—Gazprom VNIIGAZ, Razvilka, Leninsky District, Moscow Region 142717, Russia; n_buznikov@mail.ru

[2] Department of Magnetism and Magnetic Nanomaterials, Institute of Natural Sciences and Mathematics, Ural Federal University, Ekaterinburg 620002, Russia

[3] Department of Electricity and Electronics, Basque Country University UPV/EHU, 48940 Leioa, Spain

* Correspondence: galina@we.lc.ehu.es; Tel.: +34-9460-13237; Fax: +34-9460-13071

Received: 14 March 2019; Accepted: 9 April 2019; Published: 12 April 2019

Abstract: Intensive studies of the magnetoimpedance (MI) effect in nanostructured multilayers provide a good phenomenological basis and theoretical description for the symmetric case when top and bottom layers of ferromagnet/conductor/ferromagnet structure have the same thickness and consist of one magnetic layer each. At the same time, there is no model to describe the MI response in multilayered films. Here, we propose the corresponding model and analyze the influence of the multilayer parameters on the field and frequency dependences of the MI. The approach is based on the calculation of the field distribution within the multilayer by means of a solution of linerializied Maxwell equations together with the Landau–Lifshitz equation for the magnetization motion. The theoretical model developed allows one to explain qualitatively the main features of the MI effect in multilayers and could be useful for optimization of the film parameters. It might also be useful as a model case for the development of MI magnetic biosensors for magnetic biomarker detection.

Keywords: magnetic multilayers; magnetoimpedance; modeling; magnetic sensors; magnetic biosensors

---

1. Introduction

The magnetoimpedance (MI) effect implies a strong dependence of the complex impedance of a ferromagnetic conductor on an external magnetic field [1,2]. Since its rediscovery [3–6], the effect has attracted much attention due to its remarkable advantages for the development of high-sensitive magnetic field detectors [7,8]. The origin of MI can be explained in the framework of the classical skin effect, i.e., the tendency of an alternating electric current to be distributed within a conductor in such a way that the current density is largest near the conductor surface. The MI is related to the changes in the skin depth with the permeability of the ferromagnetic conductor and is observed in soft magnetic materials, which exhibit variation in permeability at low external magnetic fields. The effect was studied in detail in different magnetic materials, in particular, in amorphous wires and ribbons, electroplated wires, glass-coated microwires, and thin-film based systems.

Maximum magnitudes of the impedance variation and field sensitivity were obtained in Co-based amorphous wires and glass-coated microwires. However, for sensor miniaturization, thin-film structures could be more attractive materials. Experimental studies of the MI effect have been performed in films with different structures, such as single-layer ferromagnetic films [9,10] and three-layered films called "MI sandwich" [10–14] (see Figure 1a). Changes in the film impedance with the external magnetic field become high when the skin penetration depth reaches the order of the film thickness. For a soft magnetic film with a thickness of 1 μm, this condition is valid within the gigahertz frequency range [12]. To observe large changes in the impedance at moderate frequencies, three-layered film structures consisting of two soft magnetic films separated by a non-magnetic layer were proposed, designed, tested, and described by appropriate models [12,14]. Typical material for soft magnetic layers is nanostructured permalloy, and highly conductive Cu, Al, or Au are used for the central layer material. Thin-film sensitive elements with thicknesses of the order of microns are required for many sensor applications, including MI [13,15].

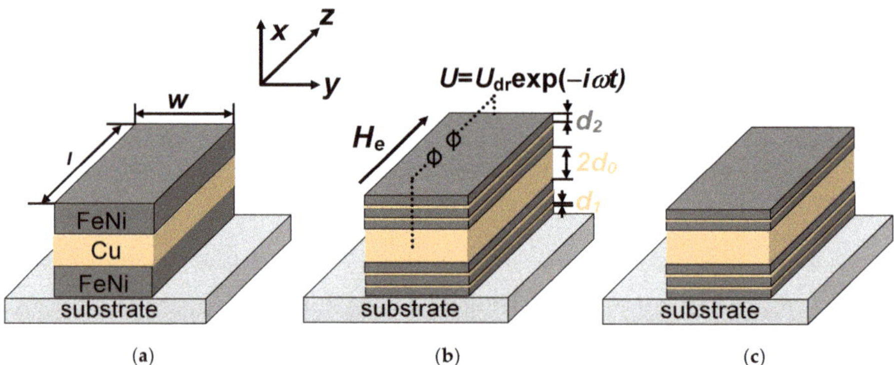

Figure 1. Schematic representation of the MI multilayered sensitive elements. (**a**) Classic "MI sandwich" without nanostructuring of magnetic layers. (**b**) MI symmetric multilayer with the same total thickness of FeNi top and bottom layers, both deposited as multilayers with Cu spacers. (**c**) MI non-symmetric multilayer with different total thicknesses of top and bottom layers: top multilayer contains two FeNi sub-layers and bottom multilayer contains three FeNi sub-layers in this particular case.

High field sensitivity of the impedance in film structures can be obtained when permalloy layers have low coercivity, high permeability, and well defined in-plane magnetic anisotropy with low local anisotropy axes distribution. However, the out-of-plane component of the anisotropy can appear in permalloy films, when the film thickness increases above the critical thickness of the transition into a "transcritical state" [16–18]. The value of the thickness corresponding to the transition depends on many parameters, including the working gas pressure during sputtering deposition [17]. This value can vary in the range from a few nanometers to a few hundred nanometers [17,19]. The appearance of the out-of-plane component of the anisotropy is ascribed to columnar structure formation, magnetocrystalline, and/or magnetoelastic anisotropy [19]. As a result, degradation of the soft magnetic properties takes place due to the transition into the "transcritical state" [16,19]. To avoid the "transcritical state" transition and to increase the total thickness of soft magnetic layers, nanostructured multilayers have been proposed and developed [20,21].

The influence of different parameters, such as the thickness of magnetic layers, material, and thickness of non-magnetic interfaces, on the magnetic properties and the MI response in multilayers was studied experimentally [22–28]. It was also demonstrated that MI in nanostructured multilayered films could be promising for the development of magnetic biosensors [29,30]. In a magnetic biosensor, non-uniform magnetic fields having a complex configuration should be analyzed. In this connection, recently, non-symmetric nanostructured multilayered films have attracted considerable attention [31,32]. Non-symmetric films were obtained by the deposition of top and bottom ferromagnetic parts of a multilayer with different thicknesses. It was found that the symmetric multilayered films have the highest field sensitivity. At the same time, non-symmetric multilayers allow one to obtain a higher MI response at high frequencies [32].

Although the MI effect in nanostructured multilayers was intensively studied in experiments, to the best of our knowledge, there is no model to describe the MI response in multilayered films. In this paper, we propose the corresponding model and analyze the influence of the multilayer parameters on the field and frequency dependences of the MI. The approach is based on the calculation of the field distribution within the multilayered film by means of a simultaneous solution of linerailzied Maxwell equations and the Landau–Lifshitz equation for the magnetization motion. Both the symmetric and non-symmetric nanostructured multilayered films are studied. The model developed allows one to explain qualitatively the main features of the MI effect in multilayers and could be useful for optimization of the film parameters.

2. Model

Let us consider a film structure, $[F/X]_m/F/C/[F/X]_n/F$, shown schematically in Figure 1. The structure consists of a highly conductive non-magnetic central layer, C, of a thickness, $2d_0$, and two external (top and bottom) multilayers. The external multilayers contain soft magnetic layers, F, of a thickness, d_2, separated by non-magnetic layers (spacers), X, of a thickness, d_1. The corresponding conductivities of the layers, C, X, and F, are σ_0, σ_1, and σ_2. Note that top and bottom multilayers may have different thicknesses, $m \neq n$. It is taken into account that materials of the central layer, C, and spacers, X, may be different [27,33]. The film structure length and width are l and w, respectively.

It is further supposed that all magnetic layers have the same physical properties. Usually, during the deposition of multilayered films, a constant magnetic field is applied along the short side of the film in order to induce the transverse magnetic anisotropy. To take into account this fact, we assume that the magnetic layers have uniaxial in-plane magnetic anisotropy, and the angle, ψ, of deviation of the anisotropy axis from the transverse direction is relatively small.

We also assume that the value of the permeability in the ferromagnetic layers is governed by the magnetization rotation only. This approximation is valid at sufficiently high frequencies (above 10 MHz), when the domain-wall motion is damped [34,35]. Furthermore, due to the averaging over the domain structure, the permeability tensor has a quasi-diagonal form. In this case, the MI of the multilayered film depends on the value of the transverse permeability only.

The alternating driving voltage, $U = U_{dr}\exp(-i\omega t)$ (where ω is the angular frequency, t is the time, and i is the imaginary unit), is applied to the multilayered structure, and the external magnetic field, H_e, is parallel to the long side of the film (see Figure 1b).

Let us restrict our consideration by the case of not too high frequencies when $\omega l/c << 1$, where c is the speed of light in vacuum. Then, the field distribution in the film can be considered to be independent of the coordinate, z. For the film length, $l = 1$ cm, this approximation is valid at frequencies, $f = \omega/2\pi <<$ 5 GHz. Moreover, since the film width, w, is much higher than its thickness, neglecting edge effects, we can suppose that the electromagnetic fields depend only on the coordinate, x, perpendicular to the film plane. This approach is adequate when the film width exceeds some critical value. This critical value, λ, depends on the layer thicknesses and static permeability in the magnetic layers [36–38]. Estimations show that $\lambda \approx 10$ μm for typical parameters of the multilayered films studied below.

In the one-dimensional approximation, the amplitudes of the longitudinal electric field, e_0, and the transverse magnetic field, h_0, in the central non-magnetic layer, C, $-d_0 < x < d_0$, satisfy Maxwell equations centimeter-gram-second (cgs) system of units is used:

$$-\frac{de_0}{dx} = (i\omega/c)h_0 ,$$
$$\frac{dh_0}{dx} = (4\pi\sigma_0/c)e_0 . \tag{1}$$

The solution of Equation (1) can be expressed as follows:

$$e_0 = (cp_0/4\pi\sigma_0)[A_0\cosh(p_0 x) + B_0\sinh(p_0 x)] ,$$
$$h_0 = A_0\sinh(p_0 x) + B_0\cosh(p_0 x) . \tag{2}$$

where, A_0 and B_0 are the constants, $p_0 = (1 - i)/\delta_0$ and $\delta_0 = c/(2\pi\omega\sigma_0)^{1/2}$. Note that for the symmetric film, $m = n$, the constant, B_0, is equal to zero.

The solution of the Maxwell equations for the field amplitudes in the non-magnetic spacers, X, can be presented in the following form:

$$
\begin{aligned}
e_1^{(j)} &= (cp_1/4\pi\sigma_1)[A_1^{(j)}\cosh(p_1 x) + B_1^{(j)}\sinh(p_1 x)], \\
h_1^{(j)} &= A_1^{(j)}\sinh(p_1 x) + B_1^{(j)}\cosh(p_1 x).
\end{aligned}
\tag{3}
$$

where, $e_1^{(j)}$ and $h_1^{(j)}$ are the amplitudes of the longitudinal electric field and the transverse magnetic field, respectively; $j = 1, \ldots, m + n$ is the non-magnetic layer number; $A_1^{(j)}$ and $B_1^{(j)}$ are the constants; $p_1 = (1 - i)/\delta_1$ and $\delta_1 = c/(2\pi\omega\sigma_1)^{1/2}$.

The field amplitudes, $e_2^{(k)}$ and $h_2^{(k)}$, in the magnetic layers, F, are given by:

$$
\begin{aligned}
e_2^{(k)} &= (cp_2/4\pi\sigma_2)[A_2^{(k)}\cosh(p_2 x) + B_2^{(k)}\sinh(p_2 x)], \\
h_2^{(k)} &= A_2^{(k)}\sinh(p_2 x) + B_2^{(k)}\cosh(p_2 x).
\end{aligned}
\tag{4}
$$

where, $k = 1, \ldots, m + n + 2$ is the magnetic layer number; $A_2^{(k)}$ and $B_2^{(k)}$ are the constants; $p_2 = (1 - i)/\delta_2$, $\delta_2 = c/(2\pi\omega\mu\sigma_2)^{1/2}$ and μ is the transverse permeability.

To find the transverse permeability in the magnetic layers, we neglect the contribution of the exchange energy. More rigorous theoretical treatment requires the inclusion of the exchange-conductivity effect in the model [39,40]. However, the contribution of the exchange-conductivity effect to the MI response is relatively low within the high-frequency range. The solution of the linearized Landau–Lifshitz equation results in the following expression for the transverse permeability, μ, in the ferromagnetic layers [41]:

$$
\mu = 1 + \frac{\gamma 4\pi M(\gamma 4\pi M + \omega_1 - i\kappa\omega)\sin^2\theta}{(\gamma 4\pi M + \omega_1 - i\kappa\omega)(\omega_2 - i\kappa\omega) - \omega^2},
\tag{5}
$$

$$
\begin{aligned}
\omega_1 &= \gamma[H_a\cos^2(\theta - \psi) + H_e\sin\theta], \\
\omega_2 &= \gamma[H_a\cos\{2(\theta - \psi)\} + H_e\sin\theta].
\end{aligned}
\tag{6}
$$

where M is the saturation magnetization of the magnetic layers, γ is the gyromagnetic constant, κ is the Gilbert damping parameter, θ is the equilibrium magnetization angle, and H_a is the anisotropy field in the ferromagnetic layers.

The equilibrium magnetization angle, θ, can be found by minimizing the free energy. The free energy can be presented as a sum of the uniaxial anisotropy energy and Zeeman energy. The minimization procedure results in the following equation for the magnetization equilibrium angle, θ:

$$
H_a\sin(\theta - \psi)\cos(\theta - \psi) = H_e\cos\theta.
\tag{7}
$$

To describe the field distribution in the external regions, we use the approximate solution for the vector potential obtained previously [37,42] in the case, $d << w$, where $d = 2d_0 + (m + n)d_1 + (m + n + 2)d_2$ is the total multilayer thickness. The field amplitudes are given by:

$$
\begin{aligned}
e_s &= C_s\frac{i\omega}{c}\left[\frac{l}{2w}\log\left(\frac{R+w}{R-w}\right) - \frac{2x}{w}\arctan\left(\frac{wl}{2Rx}\right) + \frac{1}{2}\log\left(\frac{R+l}{R-l}\right)\right], \\
h_s &= -C_s\frac{4lx}{R}\left[\frac{R^2+4x^2}{4R^2x^2+l^2w^2} - \frac{1}{R^2-w^2} - \frac{1}{R^2-l^2}\right] + C_s\frac{2}{w}\arctan\left(\frac{wl}{2Rx}\right).
\end{aligned}
\tag{8}
$$

where the subscripts, $s = 3$ and $s = 4$, correspond to the bottom and top external region, respectively, C_s are the constants, and $R = (l^2 + w^2 + 4x^2)^{1/2}$. In the symmetric film, $m = n$, the distribution of the electric field in the external region is symmetrical with respect the multilayer center, and $C_3 = C_4$.

Thus, the field distribution within the $2(m + n) + 3$ layers of the film is described by Equations (2) to (4). The total number of constants in Equations (2) to (4) is equal to $4(m + n) + 6$. The constants, A_0, B_0, $A_1^{(j)}$, $B_1^{(j)}$, $A_2^{(k)}$, and $B_2^{(k)}$, can be found from the continuity conditions for the amplitudes of the electric and magnetic fields at the interfaces between different layers. In addition, we should take into account that the driving voltage with the amplitude, U_{dr}, is applied to the film region, $-t_1 < x < t_2$, where $t_1 = d_0 + nd_1 + (n + 1)d_2$ and $t_2 = d_0 + md_1 + (m + 1)d_2$. Then, the boundary conditions at the bottom surface of the film, $x = -t_1$, can be written in following form:

$$e_2^{(1)}(-t_1) = e_3(-t_1) + U_{\mathrm{dr}}/l,$$
$$h_2^{(1)}(-t_1) = h_3(-t_1).$$

(9)

Similar expressions can be found at the top surface of the film structure, $x = t_2$:

$$e_2^{(m+n+2)}(t_2) = e_4(t_2) + U_{\mathrm{dr}}/l,$$
$$h_2^{(m+n+2)}(t_2) = h_4(t_2).$$

(10)

When the field distribution is obtained, we can find the impedance, Z, of the multilayered film as a ratio of the applied driving voltage to the total current, I, flowing through the film structure:

$$Z = \frac{U_{\mathrm{dr}}}{I} = \frac{U_{\mathrm{dr}}}{w \int_{-t_1}^{t_2} \sigma(x)e(x)\mathrm{d}x} = \frac{4\pi}{cw} \times \frac{U_{\mathrm{dr}}}{h_4(t_2) - h_3(-t_1)}.$$

(11)

To describe a relative variation of the impedance, let us introduce the MI ratio, $\Delta Z/Z$, which is given by:

$$\Delta Z/Z\,(\%) = 100 \times [Z(H_e) - Z(H_0)]/Z(H_0),$$

(12)

where H_0 is the value of the external field sufficient to saturate the impedance. In the further calculations, we assume that $H_0 = 100$ Oe, which is the typical magnitude of the maximum value of the experimentally available external magnetic field [26,31].

3. Results

3.1. Influence of Multilayer Parameters on MI Response

In this subsection, we analyze the results of the modeling of the field and frequency dependences of the MI in symmetric nanostructured multilayers. Let us assume that the central layer, C, and non-magnetic spacers, X, are made of the same material and, correspondingly, $\sigma_0 = \sigma_1$. The field dependence of the MI ratio for the multilayered film calculated for different frequencies is shown in Figure 2. Note that the results are presented only for the region of the positive fields and the calculated curves are symmetrical with respect to the sign of the external magnetic field, since hysteresis effects are neglected in the framework of the model. The dependence of the MI ratio on the external field shows a typical behavior with a maximum near the anisotropy field, H_a. It follows from Figure 2 that the maximum values of the MI ratio are achieved within the frequency range from 50 to 100 MHz.

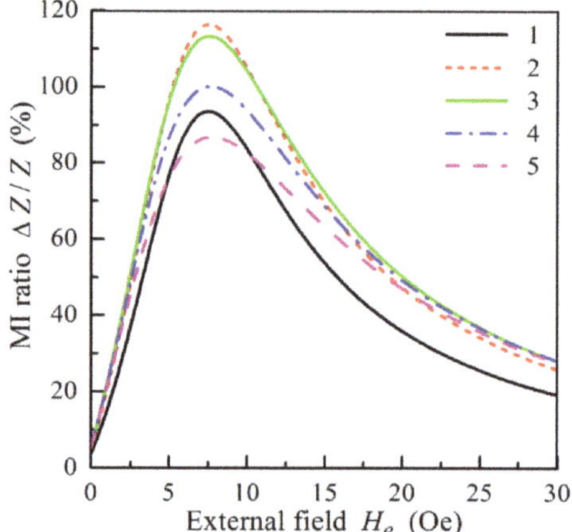

Figure 2. MI ratio, $\Delta Z/Z$, as a function of the external field, H_e, for different frequencies, f: curve 1, $f = 25$ MHz; curve 2, $f = 50$ MHz; curve 3, $f = 100$ MHz; curve 4, $f = 150$ MHz; curve 5, $f = 200$ MHz. Parameters used for calculations are $l = 1$ cm, $w = 0.02$ cm, $2d_0 = 500$ nm, $d_1 = 3$ nm, $d_2 = 100$ nm, $m = n = 4$, $M = 750$ G, $H_a = 6$ Oe, $\psi = 0.1\pi$, $\sigma_0 = \sigma_1 = 5 \times 10^{17}$ s^{-1}, $\sigma_2 = 3 \times 10^{16}$ s^{-1}, and $\kappa = 0.02$.

Figure 3 illustrates the influence of the anisotropy field, H_a, and the anisotropy axis angle, ψ, in the magnetic layers on the field dependence of the MI ratio. With a decrease of H_a, the MI ratio increases due to a growth of the transverse permeability. At the same time, the position of the peak in the field dependence of the impedance shifts towards the zero field with a decrease of the anisotropy field (see Figure 3a). As follows, from Figure 3b, the MI ratio is very sensitive to the value of the anisotropy axis angle, ψ, in the ferromagnetic layers. The MI ratio drops sharply with an increase of the deviation of the anisotropy axis from the transverse direction.

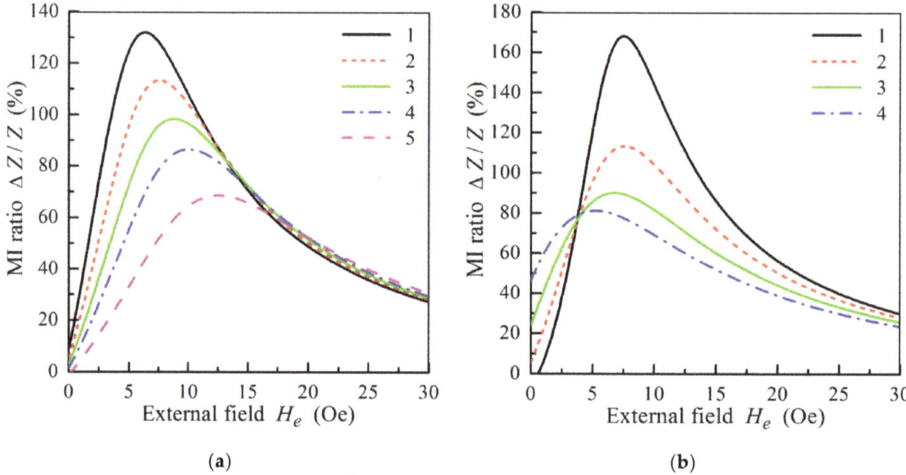

(a) (b)

Figure 3. (a) MI ratio, $\Delta Z/Z$, as a function of the external field, H_e, at $f = 100$ MHz for $\psi = 0.1\pi$ and different values of the anisotropy field, H_a: curve 1, $H_a = 5$ Oe; curve 2, $H_a = 6$ Oe; curve 3, $H_a = 7$ Oe; curve 4, $H_a = 8$ Oe; curve 5, $H_a = 10$ Oe. (b) $\Delta Z/Z$ ratio as a function of the external field, H_e, at $f = 100$ MHz for $H_a = 6$ Oe and different values of the anisotropy angle, ψ: curve 1, $\psi = 0.05\pi$; curve 2, $\psi = 0.1\pi$; curve 3, $\psi = 0.15\pi$; curve 4, $\psi = 0.2\pi$. Other parameters used for calculations are the same as in Figure 2.

Let us study the influence of the multilayer geometric parameters on the MI. For the analysis, we use the maximum MI ratio, $(\Delta Z/Z)_{max}$, which is defined as follows:

$$(\Delta Z/Z)_{max}\,(\%) = 100 \times [Z_{max} - Z(H_0)]/Z(H_0),\tag{13}$$

where Z_{max} corresponds to the peak in the field dependence of the multilayer impedance.

Figure 4a shows the frequency dependence of the maximum MI ratio, $(\Delta Z/Z)_{max}$, calculated for multilayered films with different thicknesses, $2d_0$, of the central layer. The value of $(\Delta Z/Z)_{max}$ decreases with the thickness of the central layer, and the peak in the frequency dependence of $(\Delta Z/Z)_{max}$ shifts towards higher frequencies with a decrease of $2d_0$. The results obtained are in qualitative agreement with the experimental data [24] and the results of simulation by means of the finite element method [27].

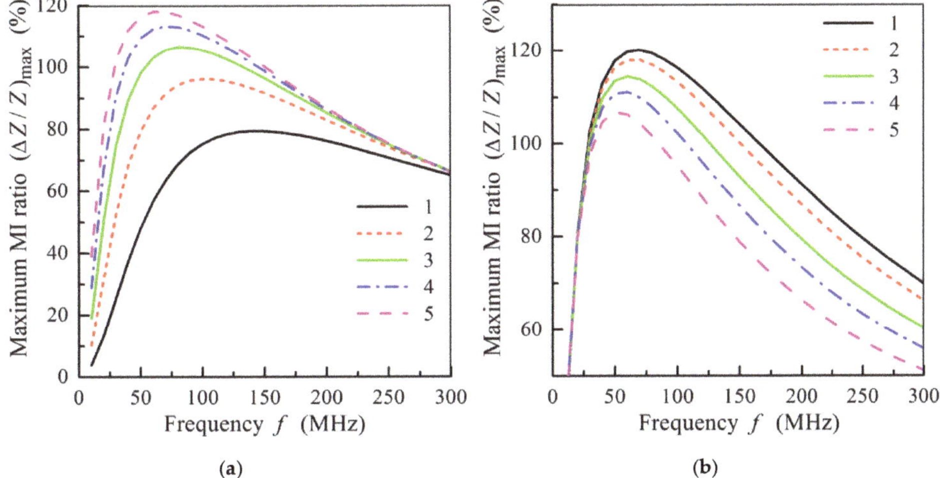

Figure 4. (**a**) Frequency dependence of the maximum MI ratio, $(\Delta Z/Z)_{max}$, at different values of the central layer thickness, $2d_0$: curve 1, $2d_0$ = 100 nm; curve 2, $2d_0$ = 200 nm; curve 3, $2d_0$ = 300 nm; curve 4, $2d_0$ = 400 nm; curve 5, $2d_0$ = 500 nm. (**b**) Frequency dependence of the maximum MI ratio, $(\Delta Z/Z)_{max}$, at different values of the thickness, d_1, of spacers: curve 1, d_1 = 2 nm; curve 2, d_1 = 3 nm; curve 3, d_1 = 5 nm; curve 4, d_1 = 7 nm; curve 5, d_1 = 10 nm. Other parameters used for calculations are the same as in Figure 2.

Figure 4b presents the effect of the thickness, d_1, of separating layers on the frequency dependence of $(\Delta Z/Z)_{max}$. Maximal values of $(\Delta Z/Z)_{max}$ are attained at low d_1, i.e., the increase of the thickness of non-magnetic spacers results in a decrease of the MI. It should be noted, however, that at low values of d_1, the exchange interactions between magnetic layers appear, which can essentially influence the MI response. The critical thickness of the non-magnetic separating layer depends significantly on the properties of the magnetic layers.

The effect of the number of magnetic layers on the frequency dependence of $(\Delta Z/Z)_{max}$ is shown in Figure 5. Note that the total thickness of the magnetic layers is constant for all films used for calculation. It follows from Figure 5 that the value of $(\Delta Z/Z)_{max}$ drops with the increase of the number of magnetic layers and with the corresponding decrease of the magnetic layer thickness.

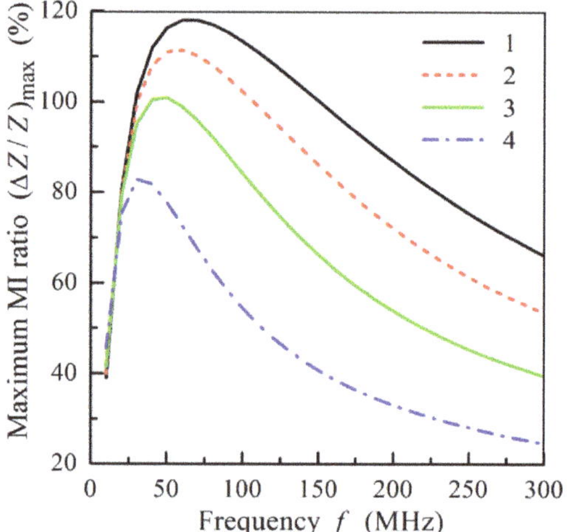

Figure 5. Frequency dependence of the maximum MI ratio, $(\Delta Z/Z)_{\max}$, for different symmetric film structures, $m = n$: curve 1, $m = 4$ and $d_2 = 100$ nm; curve 2, $m = 9$ and $d_2 = 50$ nm; curve 3, $m = 19$ and $d_2 = 25$ nm; curve 4, $m = 49$ and $d_2 = 10$ nm. Other parameters used for calculations are the same as in Figure 2.

It should be noted that the magnetic properties of the magnetic layers may change with the thickness of layers and this fact may affect significantly the MI response. In particular, an experimental study [43] showed that multilayers composed with permalloy layers of a thickness of 50 and 100 nm exhibit a similar MI ratio, whereas multilayers with thinner magnetic layers have a lower MI response. An opposite tendency was observed in another experiment [44], where it was found that the film structures with magnetic layers of a thickness of 25 nm exhibit a much higher MI ratio than films with magnetic layers with a thickness of 170 nm. This disagreement between the experimental data [44] and results of the modeling may be due to the fact that soft magnetic properties degrade in films with thick layers as a result of an approximation toward the transition into the "transcritical state" [17]. Mixed interfaces can also contribute to the balance. The volume corresponding to the interfaces is similar for multilayers with different thicknesses of magnetic layers, but the ratio between the total volume and the volume corresponding to the interfaces is different for thin and thick layers. Another contribution may come from the difference in the roughness of the interfaces corresponding to multilayers with different thicknesses of magnetic layers.

As mentioned above, materials of the central layer and non-magnetic spacers may be different. Figure 6 shows the influence of the difference in the conductivity of the central layer and spacers on the MI. It follows from Figure 6a that the value of $(\Delta Z/Z)_{\max}$ increases with a decrease of the conductivity, σ_1, of the separating layers. Note that the values of $\sigma_1 = 5 \times 10^{17} \text{s}^{-1}$ and $\sigma_1 = 5 \times 10^{16} \text{ s}^{-1}$ correspond approximately to the conductivity of copper and titanium. As follows from Figure 6b, replacing copper with titanium has a more significant effect on the MI response, when the thickness of the spacers decreases.

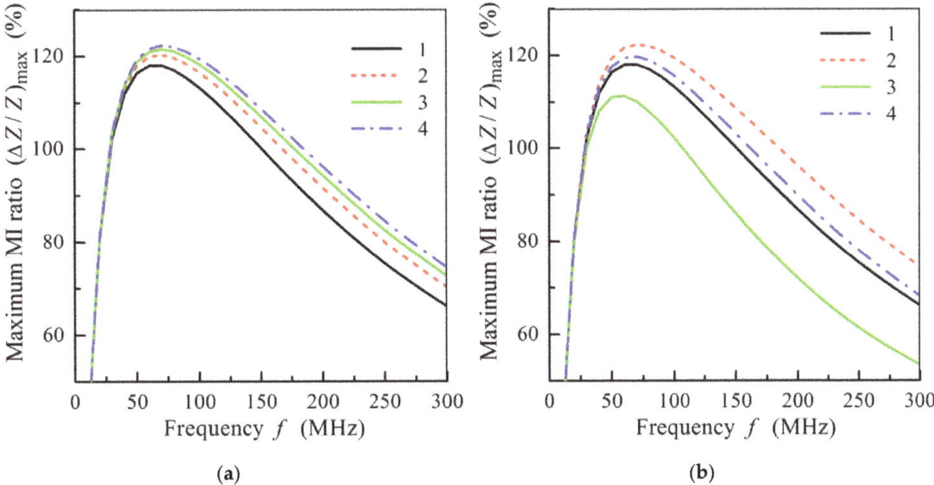

Figure 6. (**a**) Frequency dependence of the maximum MI ratio, $(\Delta Z/Z)_{\max}$, for $m = 4$, $d_2 = 100$ nm, and different values of the conductivity, σ_1, of spacers: curve 1, $\sigma_1 = 5 \times 10^{17}$ s^{-1}; curve 2, $\sigma_1 = 2 \times 10^{17}$ s^{-1}; curve 3, $\sigma_1 = 10^{17}$ s^{-1}; curve 4, $\sigma_1 = 5 \times 10^{16}$ s^{-1}. (**b**) Frequency dependence of the maximum MI ratio, $(\Delta Z/Z)_{\max}$, at different values of the conductivity, σ_1, and thickness, d_1, of spacers: curve 1, $m = 4$, $d_2 = 100$ nm, and $\sigma_1 = 5 \times 10^{17}$ s^{-1}; curve 2, $m = 4$, $d_2 = 100$ nm, and $\sigma_1 = 5 \times 10^{16}$ s^{-1}; curve 3, $m = 9$, $d_2 = 50$ nm, and $\sigma_1 = 5 \times 10^{17}$ s^{-1}; curve 4, $m = 9$, $d_2 = 50$ nm, and $\sigma_1 = 5 \times 10^{16}$ s^{-1}. Other parameters used for calculations are the same as in Figure 2.

3.2. MI in Non-Symmetric Nanostructured Multilayers

Let us now study the MI effect in non-symmetric multilayered structures. The frequency dependence of the maximum MI ratio, $(\Delta Z/Z)_{\max}$, calculated for film structures with different numbers of layers is shown in Figure 7. It is assumed that the properties of the magnetic layers are the same for the symmetric, $n = m$, and non-symmetric films, $n < m$. It follows from Figure 7a that the value of $(\Delta Z/Z)_{\max}$ decreases with the number of layers, n. The frequency of the peak in $(\Delta Z/Z)_{\max}$ increases with the growth of asymmetry between the top and bottom layers. Within the frequency range, $f > 250$ MHz, non-symmetric multilayered films exhibit a higher MI effect.

Figure 7b presents the results of calculations of $(\Delta Z/Z)_{\max}$ for the films with thinner magnetic layers. In this case, the decrease in $(\Delta Z/Z)_{\max}$ for the non-symmetric films is less pronounced in comparison with the symmetric film structure, $n = 9$. At the frequencies of the order of 150 MHz and higher the symmetric and non-symmetric films show very similar magnitudes of $(\Delta Z/Z)_{\max}$ (Figure 7b). For practical purposes, a lower working frequency may have higher importance in comparison with the maximum effect value. From this point of view, the result corresponding to $n = 9$ is very interesting as the $(\Delta Z/Z)_{\max}$ peak appears at a lower frequency in comparison with the $n = 4$ curve (Figure 7a).

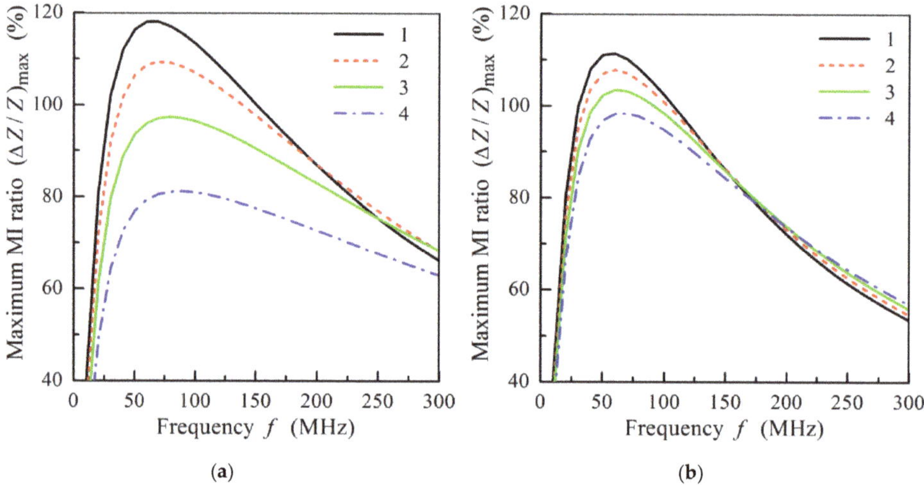

Figure 7. (**a**) Frequency dependence of the maximum MI ratio, $(\Delta Z/Z)_{max}$, at $m = 4$, $d_2 = 100$ nm, and different values of n: curve 1, $n = 4$; curve 2, $n = 3$; curve 3, $n = 2$; curve 4, $n = 1$. (**b**) Frequency dependence of the maximum MI ratio, $(\Delta Z/Z)_{max}$, at $m = 9$, $d_2 = 50$nm, and different values of n: curve 1, $n = 9$; curve 2, $n = 8$; curve 3, $n = 7$; curve 4, $n = 6$. Other parameters used for calculations are the same as in Figure 2.

4. Discussion

The aim of the work was to develop a model and theoretical analysis of MI in multilayer films with nanostructured magnetic layers. Up to now, such a model description was absent in the research literature. The results of modeling are in qualitative agreement with experimental studies of the MI in non-symmetric multilayers [31,32,44]. However, in the experiments, the change in the frequency of the peak in $(\Delta Z/Z)_{max}$ is more pronounced, when the difference in the thickness of the top and bottom layers increases. Moreover, it was found that the frequency dependence of $(\Delta Z/Z)_{max}$ differs significantly for the multilayers with odd and even numbers, n, in the bottom layer [32]. These facts clearly indicate that magnetostatic interactions between magnetic layers affect significantly the MI in non-symmetric multilayers. In fact, in all previous designs of the multilayers, the main priority was given to the evaluation of the coercivity of single-layer or three-layered structures. For example, the interaction between two magnetic layers separated by a weakly magnetic layer for $Fe_{19}Ni_{81}/Cr/Fe_{19}Ni_{81}$ and $Fe_{15}Co_{20}Ni_{65}/Cr/Fe_{15}Co_{20}Ni_{65}$ was studied [45]. The experimental data on the coercivity, domain structure parameters, and microstructure as well as the theoretical estimates showed that an increase of the thickness of the Cr can lead to a replacement of the exchange interaction between the ferromagnetic layers by the magnetostatic interaction. In its turn, the effectiveness of the magnetostatic interaction can be governed by surface defects and the layer magnetization ripple structure [45]. These conclusions correspond to the simplest symmetric structure, F/X/F, even without a central conductor, but the results for F/X/F/X/F, F/X/F/X/F/X/F, etc. are absent in the literature.

One of the weak points in a comparison of the magnetic properties of multilayered structures is a well known dependence of the properties of thin films on the preparation conditions and even particular equipment [46,47]. In combination with the strong dependence of the interaction between two magnetic layers on the thickness of the non-magnetic spacer, varying at nanoscale, the comparison becomes a very difficult task. With respect to MI multilayers, the experimental data on the adjustment thickness of the spacers are very limited. The microstructure and magnetic properties of FeNi films and FeNi (170 nm)/M/FeNi (170 nm) (M = Co, Fe, Gd, Gd-Co) multilayers were studied in [48]. In contrast to the Co and Fe spacers, Gd and Gd-Co magnetic spacers improved the softness of the FeNi/X/FeNi multilayers. The MI responses were also measured, and the highest MI variation was observed for the [FeNi/Gd (2 nm)]$_2$/FeNi case. The thickness for the minimum of coercivity in the case of the Gd spacer was 3 ± 1 nm, which is almost the accuracy limit for the sputtering technique. Although the MI measurements were performed for [FeNi/Gd (2 nm)]$_2$/FeNi multilayers, the behavior of the system for different spacer thicknesses in the cases of different numbers of magnetic layers was not studied systematically. All this means that we still need to make experimental efforts in order to improve the phenomenological basis for the next step of the theoretical development of the problem of MI multilayers.

In order to understand the role of the magnetostatic interactions, a more systematic investigation is required. In the framework of the model proposed, the magnetostatic interactions can be taken into account by introducing an additional effective field acting on the bottom layer of the film structure. Although this approach simplifies the real field distribution, it allows one to describe qualitatively the influence of the magnetostatic field on the MI in non-symmetric multilayers.

We would like to now return to the concept of the magnetic MI biosensor. Why is the problem of symmetric or non-symmetric MI structures so important in this particular case? There are two main different types of biomedical requests: Analysis of electric and magnetic properties of living systems, reflecting their functionality, and analysis of specific properties of the biocomponents. In the present work, we refer to both the first and the second kind. The first case was already demonstrated as useful devices for biomagnetic level magnetic field recording (magnetocardiogram or magnetoencefalogramm) [49] or a very first attempt to use MI detectors for diagnostics of vascular problems near stenosis [50].

The second kind are magnetic marker detecting compact analytical devices [29,30,51]. As the main principle of magnetic marker detection is an evaluation of the sum of the stray fields of all magnetizable markers [52], their conjunction can be viewed as an additional layer with particular properties (Figure 8). Magnetic markers for biomagnetic detection are spherical superparamagnetic nanoparticles or polymer composites containing spherical superparamagnetic nanoparticles, usually biocopatible iron oxides [53,54]. In the ideal case, they are all identical and carry the same magnetic moments, $\vec{m}$, in a certain applied magnetic field. External field and magnetic moments of individual markers are parallel to each other and therefore each marker creates stray fields. The measured difference between the sensor output in the absence of the magnetic markers and in their presence allows (in the calibrated system) a calculation of the amount of magnetic markers and therefore the biocomponents of interest [55,56]. One therefore treats the study of the comparison of symmetric and non-symmetric cases for MI multilayers as a model approach for improving the MI biodetector sensitivity.

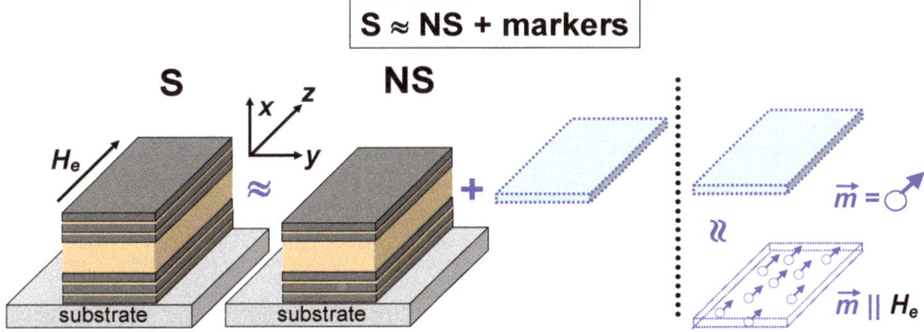

Figure 8. Schematic representation of the symmetric (S) and non-symmetric (NS) MI multilayered sensitive elements. In a model case (to mimic the layer of randomly distributed magnetic markers), the top multilayer can be substituted by a set of superparamagnetic spherical markers with the same individual moments, $\vec{m}$, oriented in the direction of the applied field, H_e.

A theoretical study of the MI in symmetric and non-symmetric nanostructured multilayers can be useful for the development of planar detectors of very low magnetic fields (of the order of biomagnetic responses) of both types described above. Of course, proper comparison of the experimental results and specially designed MI multilayers with nanostructured magnetic layers is desirable and we are now in the process of obtaining the set of required multilayered samples for comparison.

5. Conclusions

The MI effect in symmetric and non-symmetric multilayers with nanostructured magnetic layers was studied theoretically in order to analyze the influence of the multilayer parameters on the field and frequency dependences of the MI response. The proposed approach consists of a calculation of the field distribution within the multilayered film by means of a simultaneous solution of lineralizied Maxwell equations and the Landau–Lifshitz equation. The model developed allows one to explain qualitatively the main features of the MI effect in multilayers. It can be useful for optimization of the MI film parameters. It might also be useful as a model case for the development of MI magnetic biosensors for magnetic marker detection.

Author Contributions: Conceptualization, N.A.B. and G.V.K.; Funding acquisition, G.V.K.; Investigation, N.A.B. and G.V.K.; Methodology, N.A.B. and G.V.K.; Writing—original draft, N.A.B. and G.V.K. Authors are contributed equally to the research and manuscript preparation. All authors read and approved the final version of the manuscript.

Funding: This research was funded by the Russian Science Foundation, grant number 18-19-00090.

Conflicts of Interest: The authors declare no conflict of interest.

References

1. Harrison, E.P.; Turney, G.L.; Rowe, H. Electrical properties of wires of high permeability. *Nature* **1935**, *135*, 961. [CrossRef]
2. Makhotkin, V.E.; Shurukhin, B.P.; Lopatin, V.A.; Marchukov, P.Y.; Levin, Y.K. Magnetic field sensors based on amorphous ribbons. *Sens. Actuators A* **1991**, *27*, 759–762. [CrossRef]
3. Beach, R.S.; Berkowitz, A.E. Giant magnetic field dependent impedance of amorphous FeCoSiB wire. *Appl. Phys. Lett.* **1994**, *64*, 3652–3654. [CrossRef]
4. Panina, L.V.; Mohri, K. Magneto-impedance effect in amorphous wires. *Appl. Phys. Lett.* **1994**, *65*, 1189–1191. [CrossRef]
5. Machado, F.L.A.; da Silva, B.L.; Rezende, S.M.; Martins, C.S. Giant ac magnetoresistance in the soft ferromagnet $Co_{70.4}Fe_{4.6}Si_{15}B_{10}$. *J. Appl. Phys.* **1994**, *75*, 6563–6565. [CrossRef]
6. Rao, K.V.; Humphrey, F.B.; Costa-Krämer, J.L. Very Large magneto-impedance in amorphous soft ferromagnetic wires. *J. Appl. Phys.* **1994**, *76*, 6204–6208. [CrossRef]
7. Nakayama, S.; Atsuta, S.; Shinmi, T.; Uchiyama, T. Pulse-driven magnetoimpedance sensor detection of biomagnetic fields in musculatures with spontaneous electric activity. *Bios. Bioelectr.* **2011**, *27*, 34–39. [CrossRef] [PubMed]
8. Uchiyama, T.; Mohri, K.; Honkura, Y.; Panina, L.V. Recent advances of pico-Tesla resolution magnetoimpedance sensor based on amorphous wire CMOS IC MI Sensor. *IEEE Trans. Magn.* **2012**, *48*, 3833–3839. [CrossRef]
9. Sommer, R.L.; Chien, C.L. Longitudinal and transverse magneto-impedance in amorphous $Fe_{73.5}Cu_1Nb_3Si_{13.5}B_9$ films. *Appl. Phys. Lett.* **1995**, *67*, 3346–3348. [CrossRef]
10. Xiao, S.-Q.; Liu, Y.-H.; Yan, S.-S.; Dai, Y.-Y.; Zhang, L.; Mei, L.-M. Giant magnetoimpedance and domain structure in FeCuNbSiB films and sandwiched films. *Phys. Rev. B* **2000**, *61*, 5734–5739. [CrossRef]
11. Hika, K.; Panina, L.V.; Mohri, K. Magneto-impedance in sandwich film for magnetic sensor heads. *IEEE Trans. Magn.* **1996**, *32*, 4594–4596. [CrossRef]
12. Antonov, A.S.; Gadetskii, S.N.; Granovskii, A.B.; D'yachkov, A.L.; Paramonov, V.P.; Perov, N.S.; Prokoshin, A.F.; Usov, N.A.; Lagar'kov, A.N. Giant magnetoimpedance in amorphous and nanocrystalline multilayers. *Phys. Met. Metallogr.* **1997**, *83*, 612–618.

13. Morikawa, T.; Nishibe, Y.; Yamadera, H.; Nonomura, Y.; Takeuchi, M.; Taga, Y. Giant magneto-impedance effect in layered thin films. *IEEE Trans. Magn.* **1997**, *33*, 4367–4372. [CrossRef]

14. Panina, L.V.; Mohri, K. Magneto-impedance in multilayer films. *Sens. Actuators A* **2000**, *81*, 71–77. [CrossRef]

15. Gardner, D.S.; Schrom, G.; Paillet, F.; Jamieson, B.; Karnik, T.; Borkar, S. Review of on-chip inductor structures with magnetic films. *IEEE Trans. Magn.* **2009**, *45*, 4760–4766. [CrossRef]

16. Sugita, Y.; Fujiwara, H.; Sato, T. Critical thickness and perpendicular anisotropy of evaporated permalloy films with stripe domains. *Appl. Phys. Lett.* **1967**, *10*, 229–231. [CrossRef]

17. Svalov, A.V.; Kurlyandskaya, G.V.; Hammer, H.; Savin, P.A.; Tutynina, O.I. Modification of the "transcritical" state in NiFeCuMo films produced by RF sputtering. *Tech. Phys.* **2004**, *49*, 868–871. [CrossRef]

18. Coïsson, M.; Vinai, F.; Tiberto, P.; Celegato, F. Magnetic properties of FeSiB thin films displaying stripe domains. *J. Magn. Magn. Mater.* **2009**, *321*, 806–809. [CrossRef]

19. Svalov, A.V.; Aseguinolaza, I.R.; Garcia-Arribas, A.; Orue, I.; Barandiaran, J.M.; Alonso, J.; Fernandez-Gubieda, M.L.; Kurlyandskaya, G.V. Structure and magnetic properties of thin permalloy films near the "transcritical" state. *IEEE Trans. Magn.* **2010**, *46*, 333–336. [CrossRef]

20. Kurlyandskaya, G.V.; Elbaile, L.; Alves, F.; Ahamada, B.; Barrué, R.; Svalov, A.V.; Vas'kovskiy, V.O. Domain structure and magnetization process of a giant magnetoimpedance geometry FeNi/Cu/FeNi(Cu)FeNi/Cu/FeNi sensitive element. *J. Phys.: Condens. Matter* **2004**, *16*, 6561–6568. [CrossRef]

21. Correa, M.A.; Viegas, A.D.C.; da Silva, R.B.; de Andrade, A.M.H.; Sommer, R.L. GMI in FeCuNbSiB\Cu multilayers. *Physica B* **2006**, *384*, 162–164. [CrossRef]

22. Correa, M.A.; Bohn, F.; Chesman, C.; da Silva, R.B.; Viegas, A.D.C.; Sommer, R.L. Tailoring the magnetoimpedance effect of NiFe/Ag multilayer. *J. Phys. D: Appl. Phys.* **2010**, *43*, 295004. [CrossRef]

23. Kurlyandskaya, G.V.; Svalov, A.V.; Fernández, E.; García-Arribas, A.; Barandiarán, J.M. FeNi-based magnetic layered nanostructures: Magnetic properties and giant magnetoimpedance. *J. Appl. Phys.* **2010**, *107*, 09C502. [CrossRef]

24. Volchkov, S.O.; Fernández, E.; García-Arribas, A.; Barandiaran, J.M.; Lepalovskij, V.N.; Kurlyandskaya, G.V. Magnetic properties and giant magnetoimpedance of FeNi-based nanostructured multilayers with variable thickness of the central Cu lead. *IEEE Trans. Magn.* **2011**, *47*, 3328–3331. [CrossRef]

25. Kurlyandskaya, G.V.; García-Arribas, A.; Fernández, E.; Svalov, A.V. Nanostructured magnetoimpedance multilayers. *IEEE Trans. Magn.* **2012**, *48*, 1375–1380. [CrossRef]

26. Vas'kovskii, V.O.; Savin, P.A.; Volchkov, S.O.; Lepalovskii, V.N.; Bukreev, D.A.; Buchkevich, A.A. Nanostructuring effects in soft magnetic films and film elements with magnetic impedance. *Tech. Phys.* **2013**, *58*, 105–110. [CrossRef]

27. García-Arribas, A.; Fernández, E.; Svalov, A.; Kurlyandskaya, G.V.; Barandiaran, J.M. Thin-film magneto-impedance structures with very large sensitivity. *J. Magn. Magn. Mater.* **2016**, *400*, 321–326. [CrossRef]

28. Correa, M.A.; Bohn, F. Manipulating the magnetic anisotropy and magnetization dynamics by stress: Numerical calculation and experiment. *J. Magn. Magn. Mater.* **2018**, *453*, 30–35. [CrossRef]

29. Kurlyandskaya, G.V.; Fernández, E.; Safronov, A.P.; Svalov, A.V.; Beketov, I.; Burgoa Beitia, A.; García-Arribas, A.; Blyakhman, F.A. Giant magnetoimpedance biosensor for ferrogel detection: Model system to evaluate properties of natural tissue. *Appl. Phys. Lett.* **2015**, *106*, 193702. [CrossRef]

30. Wang, T.; Zhou, Y.; Lei, C.; Luo, J.; Xie, S.; Pu, H. Magnetic impedance biosensor: A review. *Biosens. Bioelectron.* **2017**, *90*, 418–435. [CrossRef]

31. Chlenova, A.A.; Kurlyandskaya, G.V.; Volchkov, S.O.; Lepalovskij, V.N.; El Kammouni, R. Nanostructured magnetoimpedance multilayers with different thickness of FeNi components. *Solid State Phenom.* **2014**, *215*, 342–347. [CrossRef]

32. Kurlyandskaya, G.V.; Chlenova, A.A.; Fernández, E.; Lodewijk, K.J. FeNi-based flat magnetoimpedance nanostructures with open magnetic flux: New topological approaches. *J. Magn. Magn. Mater.* **2015**, *383*, 220–225. [CrossRef]

33. García-Arribas, A.; Combarro, L.; Goriena-Goikoetxea, M.; Kurlyandskaya, G.V.; Svalov, A.V.; Fernández, E.; Orue, I.; Feuchtwanger, J. Thin-film magnetoimpedance structures onto flexible substrates as deformation sensors. *IEEE Trans. Magn.* **2017**, *53*, 2000605. [CrossRef]

34. Atkinson, D.; Allwood, D.A.; Xiong, G.; Cooke, M.D.; Faulkner, C.C.; Cowburn, R.P. Magnetic domain-wall dynamics in a submicrometre ferromagnetic structure. *Nat. Mater.* **2003**, *2*, 85–87. [CrossRef] [PubMed]

35. Chen, D.; Muñoz, J.; Hernando, A.; Vázquez, M. Magnetoimpedance of metallic ferromagnetic wires. *Phys. Rev. B* **1998**, *57*, 10699–10704. [CrossRef]
36. Makhnovskiy, D.P.; Panina, L.V. Size effect on magneto-impedance in layered films. *Sens. Actuators A* **2000**, *81*, 91–94. [CrossRef]
37. Sukstanskii, A.; Korenivski, V.; Gromov, A. Impedance of a ferromagnetic sandwich strip. *J. Appl. Phys.* **2001**, *89*, 775–782. [CrossRef]
38. Panina, L.V.; Makhnovskiy, D.P.; Mapps, D.J.; Zarechnyuk, D.S. Two-dimensional analysis of magnetoimpedance in magnetic/metallic multilayers. *J. Appl. Phys.* **2001**, *89*, 7221–7223. [CrossRef]
39. Kraus, L. The theoretical limits of giant magneto-impedance. *J. Magn. Magn. Mater.* **1999**, *196–197*, 354–356. [CrossRef]
40. Ménard, D.; Yelon, A. Theory of longitudinal magnetoimpedance in wires. *J. Appl. Phys.* **2000**, *88*, 379–393. [CrossRef]
41. Panina, L.V.; Mohri, K.; Ushiyama, T.; Noda, M.; Bushida, K. Giant magneto-impedance in Co-rich amorphous wires and films. *IEEE Trans. Magn.* **1995**, *31*, 1249–1260. [CrossRef]
42. Gromov, A.; Korenivski, V.; Haviland, D.; van Dover, R.B. Analysis of current distribution in magnetic film inductors. *J. Appl. Phys.* **1999**, *85*, 5202–5204. [CrossRef]
43. Fernández, E.; Svalov, A.V.; Kurlyandskaya, G.V.; García-Arribas, A. GMI in nanostructured FeNi/Ti multilayers with different thicknesses of the magnetic layers. *IEEE Trans. Magn.* **2013**, *49*, 18–21. [CrossRef]
44. Chlenova, A.A.; Svalov, A.V.; Kurlyandskaya, G.V.; Volchkov, S.O. Magnetoimpedance of FeNi-based asymmetric sensitive elements. *J. Magn. Magn. Mater.* **2016**, *415*, 87–90. [CrossRef]
45. Vas'kovskii, V.O.; Savin, P.A.; Lepalovskij, V.N.; Ryazantsev, A.A. Multilevel interaction between layers in layered film structures. *Phys. Solid State* **1997**, *39*, 1958–1960. [CrossRef]
46. Handrich, K.; Kobe, S. *Amorphe Ferro- und Ferrimagnetika (Amorphous Ferro- and Ferrimagnets)*; Akademie-Verlag: Berlin, Germany, 1980.
47. Chikazumi, S. *Physics of Magnetism*, 2nd ed.; John Wiley: New York, NY, USA, 1997.
48. Svalov, A.V.; Fernandez, E.; Garcia-Arribas, A.; Alonso, J.; Fdez-Gubieda, M.L.; Kurlyandskaya, G.V. FeNi-based magnetoimpedance multilayers: Tailoring of the softness by magnetic spacers. *Appl. Phys. Lett.* **2012**, *100*, 162410. [CrossRef]
49. Uchiyama, T.; Nakayama, S.; Mohri, K.; Bushida, K. Biomagnetic field detection using very high sensitivity magnetoimpedance sensors for medical applications. *Phys. Status Solidi A* **2009**, *206*, 639–643. [CrossRef]
50. Volchkov, S.O.; Chlenova, A.A.; Lepalovskij, V.N. Modelling of thin film magnetoimpedance sensitive element designed for biodetection. *EPJ Web Conf.* **2018**, *185*, 10005. [CrossRef]
51. Blanc-Béguin, F.; Nabily, S.; Gieraltowski, J.; Turzo, A.; Querellou, S.; Salaun, P.Y. Cytotoxicity and GMI bio-sensor detection of maghemite nanoparticles internalized into cells. *J. Magn. Magn. Mater.* **2009**, *321*, 192–197. [CrossRef]
52. Baselt, D.R.; Lee, G.U.; Natesan, M.; Metzger, S.W.; Sheehan, P.E.; Colton, R.J. A biosensor based on magnetoresistance technology. *Biosens. Bioelectron.* **1998**, *13*, 731–739. [CrossRef]
53. Darton, N.J.; Ionescu, A.; Llandro, J. (Eds.) *Magnetic Nanoparticles in Biosensing and Medicine*; Cambridge University Press: Cambridge, UK, 2019.
54. Megens, M.; Prins, M. Magnetic biochips: A new option for sensitive diagnostics. *J. Magn. Magn. Mater.* **2005**, *293*, 702–708. [CrossRef]
55. Beato-López, J.J.; Pérez-Landazábal, J.I.; Gómez-Polo, C. Magnetic nanoparticle detection method employing non-linear magnetoimpedance effects. *J. Appl. Phys.* **2017**, *121*, 163901. [CrossRef]
56. Yang, Z.; Wang, H.; Guo, P.; Ding, Y.; Lei, C.; Luo, Y. A multi-region magnetoimpedance-based bio-analytical system for ultrasensitive simultaneous determination of cardiac biomarkers myoglobin and C-reactive protein. *Sensors* **2018**, *18*, 1765. [CrossRef]

Article

Magnetoimpedance Effect in the Ribbon-Based Patterned Soft Ferromagnetic Meander-Shaped Elements for Sensor Application

Zhen Yang [1,2], **Anna A. Chlenova** [1,3], **Elizaveta V. Golubeva** [1], **Stanislav O. Volchkov** [1], **Pengfei Guo** [2], **Sergei V. Shcherbinin** [1,4] **and Galina V. Kurlyandskaya** [1,5,*]

[1] Department of Magnetism and Magnetic Nanomaterials, Ural Federal University, 620002 Ekaterinburg, Russia; zhc025@alumni.sjtu.edu.cn (Z.Y.); anniaally@gmail.com (A.A.C.); golubeva.elizaveta.v@gmail.com (E.V.G.); stanislav.volchkov@urfu.ru (S.O.V.); scher30@yandex.ru (S.V.S.)
[2] School of Physics and Electronic Engineering, Xinyang Normal University, Xinyang 464000, China; guopengfei2010@126.com
[3] Institute of Metal Physics UD RAS, 620016 Ekaterinburg, Russia
[4] Institute of Electrophysics UD RAS, 620016 Ekaterinburg, Russia
[5] University of the Basque Country UPV-EHU, 48940 Leioa, Spain
[*] Correspondence: galina@we.lc.ehu.es; Tel.: +34-9460-13237; Fax: +34-9460-13071

Received: 7 May 2019; Accepted: 27 May 2019; Published: 29 May 2019

Abstract: Amorphous and nanocrystalline soft magnetic materials have attracted much attention in the area of sensor applications. In this work, the magnetoimpedance (MI) effect of patterned soft ferromagnetic meander-shaped sensor elements has been investigated. They were fabricated starting from the cobalt-based amorphous ribbon using the lithography technique and chemical etching. Three-turn (S1: spacing s = 50 μm, width w = 300 μm, length l = 5 mm; S2: spacing s = 50 μm, width w = 400 μm, length l = 5 mm) and six-turn (S3: s = 40 μm, w = 250 μm, length l = 5 mm; S4: s = 40 μm, w = 250 μm and l = 8 mm) meanders were designed. The 'n' shaped meander part was denominated as "one turn". The S4 meander possesses a maximum MI ratio calculated for the total impedance $\Delta Z/Z \approx 250\%$ with a sensitivity of about 36%/Oe (for the frequency of about 45 MHz), and an MI ratio calculated for the real part of the total impedance $\Delta R/R \approx 250\%$ with the sensitivity of about 32%/Oe (for the frequency of 50 MHz). Chemical etching and the length of the samples had a strong impact on the surface magnetic properties and the magnetoimpedance. A comparative analysis of the surface magnetic properties obtained by the magneto-optical Kerr technique and MI data shows that the designed ferromagnetic meander-shaped sensor elements can be recommended for high frequency sensor applications focused on the large drop analysis. Here we understand a single large drop as the water-based sample to analyze, placed onto the surface of the MI sensor element either by microsyringe (volue range 0.5–500 μL) or automatic dispenser (volume range 0.1–50 mL).

Keywords: Magnetoimpedance effect; amorphous ribbons; patterned ribbons; meander sensitive element; magnetic field sensor

---

1. Introduction

The magnetoimpedance (MI) effect is a classic electromagnetic phenomenon which can be described as a significant change of the total impedance of a ferromagnetic conducting sensor element applied to an external magnetic field when a high frequency alternating current flows through it [1–3]. The first magnetic field sensors working on the principle of the change of alternating current resistance applied to an external magnetic field were reported by Makhotkin et al. [4] even before the MI phenomenon was given its proper name [2,3].

In recent years, MI sensors have attracted special attention for their promising technological applications in different fields [5–8]. Cobalt-based amorphous ribbons show good MI characteristics with high sensitivity with respect to applied magnetic fields, sufficient even for biosensor applications [9,10]. Cobalt-based rapidly quenched materials have reasonably high saturation magnetization and nearly zero magnetostriction constant value, characteristics that make them good candidates for very different applications [11–13]. One of the shortcomings of the rapidly quenched ribbon-based materials is their relatively large size. It was shown previously that the shape anisotropy plays a crucial role and the MI ratio decreases significantly in the ribbons with a length of less than a few centimeters [14,15]. One of the solutions to such a problem was to create patterned elements in the shape of the meander [16–18] when the flat surface of the sensor element is required. Although meander geometry was previously applied with success for the case of MI multilayered structures [19–21], the proposed solution was shown to be useful in selected applications as the ribbon patterning technology is cheaper in comparison with the thin films case, despite the disadvantage of a lower degree of compatibility with semiconductor electronics, and the interaction effect resulting in a decrease of the overall sensitivity. The effective length in a meander-shaped MI sensor element is increased in comparison with the overall maximum dimension, providing the MI meander element a special advantage. At the same time, possible interactions between the fields created by the different parts of a meander can be complicated, and this geometry requires additional investigation.

Low working frequency is usually preferred for sensor applications as it makes the electronics circuit cheaper and insures easier processing of the electronic signals. At the same time, the frequency of the exciting current is an important parameter for determining the MI value [1,12]. Further development of electronic circuits and measurement techniques have possibly extended the frequency range of MI-based sensor applications [22–24] to higher frequencies where the signal-to-noise ratio can be improved [12].

Despite the fact that MI materials with a very high MI ratio and sensitivity, combined with a small size, have already been reported [25,26], there are particular biomedical applications for which flat biosensors are required to measure liquids with a relatively large bioanalyte drop. Such applications can be found in the environmental control and technological control of reservoirs. It is worth mentioning the existence of similar solutions found earlier for giant magnetoresistance-based elements for magnetic label detection [27]. In these cases, the resulting sensor elements consisted of wound lines forming the spiral structures. They were assembled into the element consisting of tens of the elements incorporated into the area of few squared millimeters. The signal for detection was collected from a number of elements after appropriate biochemical treatment. Therefore, we proposed to design a relatively large sensor element for the evaluation of the signal from one element. In future it might be useful for testing the composition with four elements in a bridge configuration.

In this work, the magnetic properties and magnetoimpedance of patterned Co-based amorphous ribbons were comparatively analyzed. We made the first step to work out the protocol of evaluation of magnetic properties of such complex subjects. MI was measured in the extended frequency range of 1–200 MHz. The patterning was done by lithography and chemical etching in the shape of meander sensor elements with three- or six-turns, with design focused on large drop analysis.

2. Materials and Methods

2.1. Fabrication of Ribbon-Based Patterned Soft Ferromagnetic Multi-Turn Sensitive Elements

First of all, we designed and developed small sensor elements in the geometry of the meander shape close to a square, aiming to develop a particular type of detector focused on the single large drop analysis of bioanalytes. Water droplet size is very important for different areas of analysis. Here, we understand a single large drop as the water-based sample to analyze, laced onto the surface of the MI sensor element either by microsyringe (volue range 0.5–500 μL) or automatic dispenser (volume range 0.1–50 mL). The fabrication process of patterned ribbons constitutes a series of steps. Initially the as-cast ribbon was produced by the rapid quenching technique [24,25] followed by cutting, field annealing, mechanical polishing and micromachining. Amorphous ribbons in the as-cast state were obtained with $Co_{75}Fe_{10}Ni_2Si_8B_5$ nominal composition and a thickness of about 20 µm. Afterward, ribbons were cut in 4×3 cm pieces via the SPF-7100 die sawing system, and the subjected to transverse magnetic field annealing treatment in order to induce transverse magnetic anisotropy. The magnetic field during field annealing was applied perpendicular to the ribbon axis (perpendicular to the direction of the ribbon movement during the solidification). The details of the annealing process can be found elsewhere [23]. Mechanical polishing was performed by diamond sand-paper with a gradual reduction of the grain size in order to reduce the surface roughness having a negative impact on the value of MI sensitivity [28]. After polishing, the thickness of the ribbons was close to 15 µm. In order to reduce the etching error, chemical etching was carried out in a thermostatic water bath. Micromachining was done by micro electro-mechanical system (MEMS) technology, including bonding, lithography, and chemical etching. After etching, all patterned ribbons in the meander shape were polished again to ensure a smooth surface. The last request is especially important in biosensor applications for magnetic label detection.

Although different geometries obtained by the employed technique were tested, we selected just a few of them with reasonable quality (Figure 1). With the amorphous ribbons it is impossible to ensure a flat cross-section of the samples for the individual line width below 0.1 mm. Planning the creation of the magnetic marker sensor for the in-liquid detection, one must keep the surface geometry as flat as possible. Meanders with a larger number of turns are difficult to fabricate by etching; therefore we selected some simple options for testing. Figure 1a,b shows the surface features of the amorphous ribbon after both chemical etching and polishing.

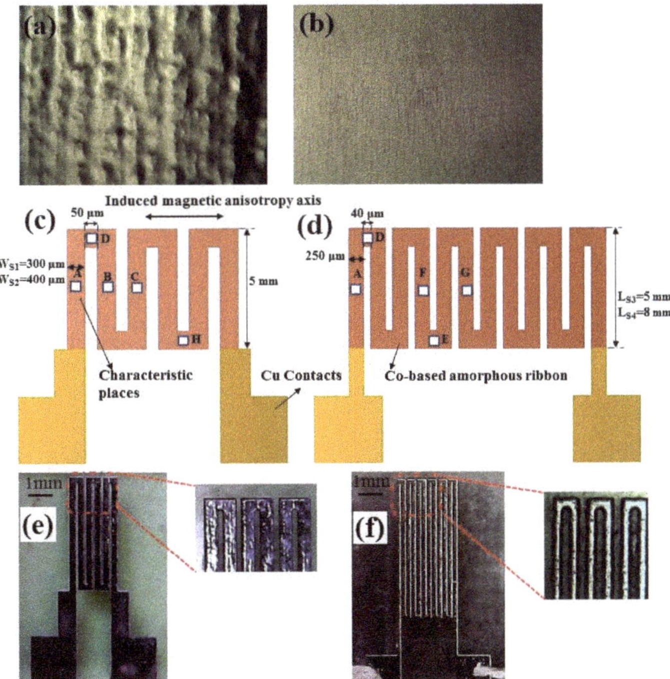

Figure 1. Surface micrograph of the $Co_{75}Fe_{10}Ni_2Si_8B_5$ ribbon after chemical etching (**a**). Surface micrograph of the chemical etched $Co_{75}Fe_{10}Ni_2Si_8B_5$ ribbon after polishing (**b**). The description of the selected types of the patterned MI meanders (**c,d**). General view of the fabricated patterned MI meanders S1 and S4 (**e,f**). Part (**c**) shows three-turn and part (**d**) shows six-turn meanders.

The micropatterned ribbon sensor elements were designed as compact n-shaped meander structures. Figure 1c,d shows selected examples of the meander geometry. The 'n' shaped meander part was denominated as a "one turn" sample.

Three-turn meanders were denominated as S1 and S2, and six-turn meanders were denominated as S3 and S4. Details of the geometries of all samples are given in Table 1. The length close to 8 mm was previously described in the literature as a critical length for non-patterned amorphous ribbon [13,14]. The spacing of 50 μm and length of 5 mm were used for S1 and S2 meanders and the same spacing of 40 μm and width of 250 μm were used for S3 and S4 meanders. Figure 1e,f shows a general view of the fabricated S1 and S4 meanders. High magnification allows an estimation of the quality of the fabricated sensitive elements.

Table 1. Selected parameters of $Co_{75}Fe_{10}Ni_2Si_8B_5$ meanders.

Sample	Number of Turns	Spacing	Width	Length
S1	3	50 μm	300 μm	5 mm
S2	3	50 μm	400 μm	5 mm
S3	6	40 μm	250 μm	5 mm
S4	6	40 μm	250 μm	8 mm

2.2. Measurement of Surface Magnetic Properties of the Samples

Surface magnetic properties of the samples were investigated by the magneto-optical Kerr effect (MOKE) using a Kerr microscope (Evico magnetics GmbH, Dresden, Germany). MOKE measurements were performed by collecting the signal in the longitudinal mode, i.e., the component of magnetization lying in the sample plane and parallel to the direction of incidence of light was analyzed. Figure 1c,d shows some selected positions—A, B, C, D, E, F, G, H for MOKE measurement (white squares) according to the symmetric structure of the meander patterning. The signal was obtained from the plot area of 0.5×0.5 mm^2. Some characteristic positions named by the same letter were picked up for the same location for different samples. The magnetic field up to 270 Oe was applied along the ribbon axis coinciding with longer part of the "n"- shaped direction. Meanwhile the magnetic field was also applied perpendicular to the ribbon axis for comparative analysis.

MOKE is a surface sensitive method that probes the surface magnetization. The origin of the high frequency alternating current impedance is connected to the change of the dynamic magnetic permeability and skin effect [1–3]. As the high frequency alternating current flows close to the surface, it is the surface anisotropy that makes a significant contribution to the MI effect value.

We made the first step to work out the protocol of evaluation of the magnetic properties of such complex subjects. On one hand, the bulk measurements with vibrating sample magnetometer (VSM) could provide more complete information because the VSM signal comes not only from the surface but also from the volume of the massive sample. On the other hand, the information obtained from the whole meander is quite mixed as the shape anisotropy and interactions between different parts of the meanders are crucial. There is no detailed evaluation of local magnetic properties of such structures in the literature and we made a step in this direction. As will be shown below, the proposed methodology seems to be working.

2.3. Measurement of MI

In order to ensure high-quality MI measurements and to minimize the "microstripe line" holder parasite signal contribution, special sample holders were prepared in accordance with the size of each sample (Figure 2b). The total effective impedance of the sample installed in a measuring circuit ("microstrip" line in this particular case) consisted of two different parts. The first one was the intrinsic impedance of the sample. It changes under the application of the external magnetic field. The second one is the external impedance attributed to the circuit itself. The last contribution does not change in the course of the external magnetic field application. This means that the measured effect is lower than the one corresponding to the intrinsic impedance of the MI sensor element.

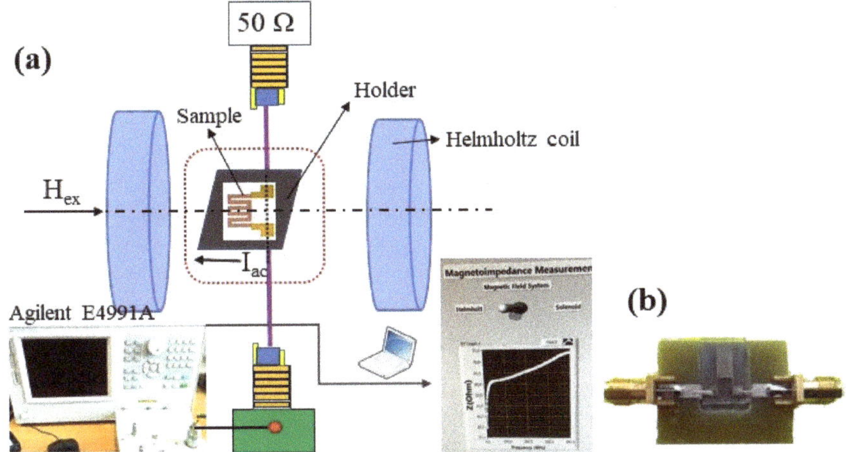

Figure 2. Schematic view of the MI measuring system for meander samples (**a**) and and "microstrype line" type sample holder with incorporated meander sample (**b**).

The impedance was measured as a function of alternating current frequency in the frequency range of 1 to 200 MHz using an automatic system based on a precision Agilent HP E4991A impedance analyzer. The calibration of the system and mathematical subtraction of the contributions of the test fixture were in accordance with a previously established procedure [29]. Total impedance (Z) was measured as a function of the external magnetic field for the exciting current amplitude of 10 mA in all measurements. The external quasi-static magnetic field (H) generated by Helmholtz coils was changed in the interval +110 to −110 Oe and aligned with the long parts of the meander.

Measurements in the decreasing field were denominated as the "down" branch; measurements in the increasing field were denominated as the "up" branch. The MI ratio for total impedance variation and its real part in the external magnetic field were calculated as follows:

$$\Delta Z/Z = 100 \times (Z(H) - Z(H_{max}))/Z(H_{max}) \tag{1}$$

$$\Delta R/R = 100 \times (R(H) - R(H_{max}))/R(H_{max}) \tag{2}$$

where H_{max} = 110 Oe. The maximum ratio at each frequency was denoted $\Delta Z/Z_{max}$ for the total impedance and $\Delta R/R_{max}$ for its real part. An important characteristic of the MI response is the maximal sensitivity with respect to the external magnetic field. The MI sensitivity with respect to the applied magnetic field was calculated as follows:

$$S(\Delta Z/Z) = \Delta(\Delta Z/Z)/\Delta H \tag{3}$$

$$S(\Delta R/R) = \Delta(\Delta R/R)/\Delta H \tag{4}$$

where $\Delta H \approx 0.1$ Oe is the field increment for calculation of ΔZ and ΔR.

3. Results and Discussion

The quasi-static magnetic properties of the meanders were measured by the MOKE microscope. Hysteresis loops taken from the characteristic sections of the samples (Figure 1) are shown in the Figure 3. The hysteresis loops in point A have the small coercive force and high magnetic permeability, which are common for amorphous magnetically soft ribbons [10,29]. More complex types of magnetization reversal were observed for the non-external long sections of the meander due to the complicated process of magnetization interactions with the adjacent parts of the element. Although, one can notice a great difference between the magnetization processes in points B and C of the samples S1, S2 and S3. In the case of the sample S4, only a slight increase in the coercive force takes place. As this sample is longer than the others, one can suppose that the changes in the type of the hysteresis loops are caused by an influence of the smaller contribution of the meanders' round sections (as points D and H in Figure 2) as well as due to the stronger contribution of the demagnetizing factor of the elongated parts. Although the details of such a mechanism are not clear, the origin can be explained by the presence of the stray fields of different meanders' areas and surface, including the border defects that appeared during the samples' production process. At the same time, in order to obtain statistics and better insight, M(H) loops were measured many times at the same position and allow a small displacement. As the M(H) loops in these conditions did not change very much we have concluded that border effects were not crucial.

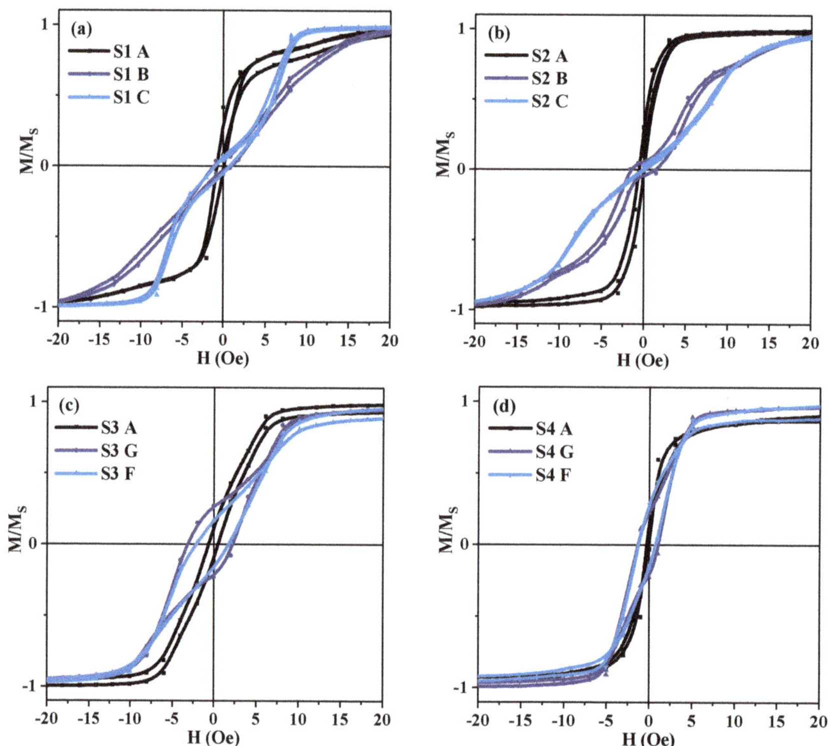

Figure 3. MOKE hysteresis loops taken from the corresponding characteristic sections A, B (G), C (F) of different $Co_{75}Fe_{10}Ni_2Si_8B_5$ meander samples. The external magnetic field is applied parallel to the long side of the meanders and perpendicular to the induced magnetic anisotropy axis.

Also, one can notice that the magnetic properties of the turning sections of meanders (points D, E, H) differ from the properties of the middle sections (Figures 3 and 4). In these regions the material of the meander reaches a saturated state only in very high magnetic fields above 200 Oe. Moreover, in the case of the sample S4 a certain increase in the coercive force is observed. This fact supports our hypothesis about role of the shape, and to a lesser extent the defects caused by the etching on local magnetic properties and effective magnetic anisotropy axis dispersion.

Although point-by-point analysis of the local hysteresis loops is useful, one can also compare the whole set of the obtained shapes of M(H) loops. It is clear that in the S4 case, M(H) loops for very different points differ from each other to a lesser extent, i.e., the magnetic material shows a lower degree of inhomogeneity in the sense of the dominating magnetization mechanisms. A similar concept was previously discussed in the literature for non-patterned amorphous and nanocrystalline ribbon, we refer to the role of the anisotropy distribution in the formation of the MI response [30,31].

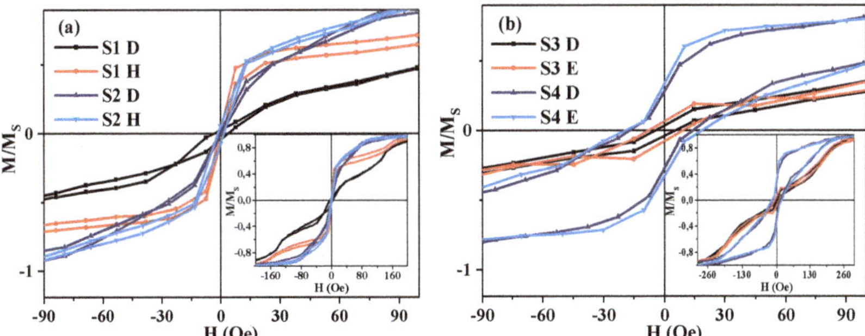

Figure 4. MOKE hysteresis loops taken from the same turning sections of the three-turn (**a**) and six-turn (**b**) $Co_{75}Fe_{10}Ni_2Si_8B_5$ meander samples.

As it can be seen (Figures 5 and 6), complex quasistatic magnetic properties of the meander samples also reflect their MI responses. One can notice several field intervals with different curvature of the MI curves, i.e., behavior, which is usually not observed for simple amorphous ribbon quenched samples [10] in the shape of a single elongated stripe. These intervals are marked as A, B and C (Figure 6) characteristic areas.

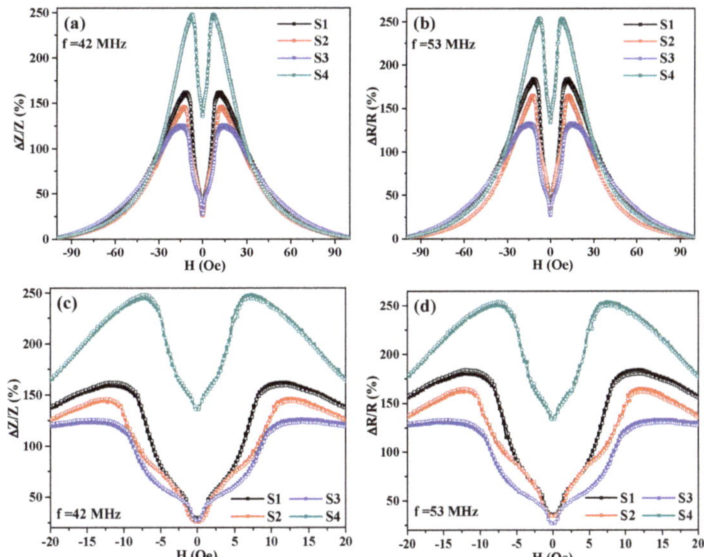

Figure 5. Field dependence of the MI ratio for the total impedance (**a,c**) and its real part (**b,d**) for the $Co_{75}Fe_{10}Ni_2Si_8B_5$ meander samples; (**c,d**)—the same dependencies in a small field range. "Down" curves for the magnetic field change from 100 Oe to −100 Oe are marked with solid symbols, "Up" curves for the magnetic field change from −100 Oe to 100 Oe are marked with empty symbols.

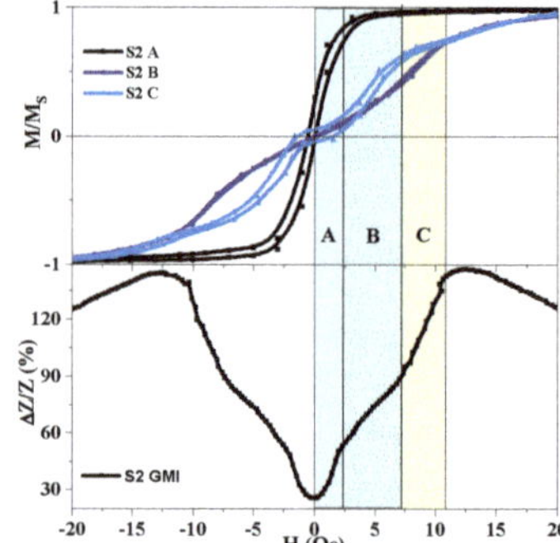

Figure 6. Comparison of the magnetization curves and field dependence of the MI ratio for total impedance of the three-turns $Co_{75}Fe_{10}Ni_2Si_8B_5$ meander S2. Characteristic field intervals with different curvature of the GMI curves are marked as A, B and C.

In small magnetic fields (interval A) the border parts of the meanders (point A) quickly magnetize, meanwhile the middle sections keep staying almost demagnetized and hence do not contribute significantly to the MI response. In the higher fields (intervals B and C) the border sections become saturated and stop contributing to the MI. The second change of curvature of M(H) curves is observed for both points B and C. Although the explanation of the origin of such behavior requires an additional investigation of the position of about 10 Oe, for both M(H) and MI, the curves indicate a strong contribution of the anisotropy field.

Despite the complexity of the dynamic magnetization process described above the meander samples used in this work show high MI response and its sensitivity to an external magnetic field (Figures 7 and 8). As is shown in Figure 8, the S4 meander demonstrates the highest values for both parameters: GMI response $\Delta Z/Z = 244\%$ and its sensitivity $S(\Delta Z/Z) = 35\%/Oe$. It also correlates with Figure 3, which shows that there is only a slight difference in magnetization curves for the different sample areas. As expected, the highest MI effect and its sensitivity are observed for the case where the most uniform anisotropy is insured. Both effects can be explained by a lower anisotropy axis dispersion in comparison with the other samples because of its longer length, and hence a smaller impact of the turning sections or the shape anisotropy.

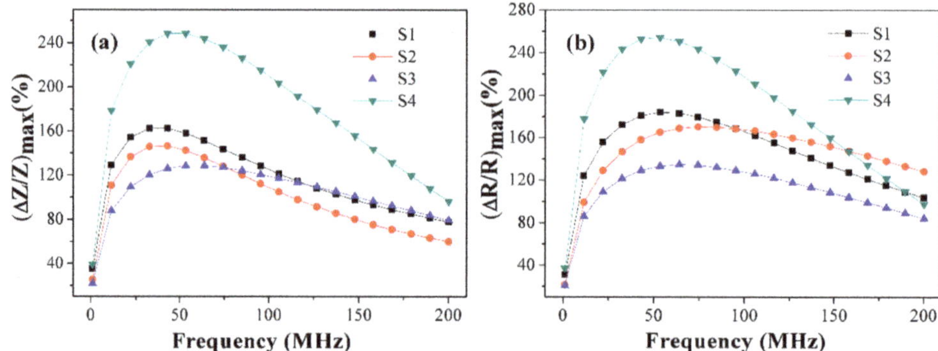

Figure 7. Frequency dependence of the maximal MI ratio for the (**a**) total impedance and (**b**) its real part for $Co_{75}Fe_{10}Ni_2Si_8B_5$ meander samples.

The frequency dependences of the maximum of the MI ratio for total impedance and the real part of the total impedance are shown in Figure 7. The maximum $\Delta Z/Z_{max}$ ratio of about 250% appeared at a frequency of 42 MHz for the six-turn sample S4 (Figure 7a). However, the $\Delta Z/Z_{max}$ ratios (162% for S1 and 146% for S2) at a lower frequency of 32 MHz was larger than that of the six-turn sample S3 (129%). The main reason may be that S1 and S2 samples had a better proportion of width and spacing. Compared with the $\Delta Z/Z_{max}$ ratios of the three-turn samples S1 and S2, we can find that the MI ratio of the sample S1 with a narrow line-width was larger than that of sample S2. These results were consistent with the earlier reports [26,30].

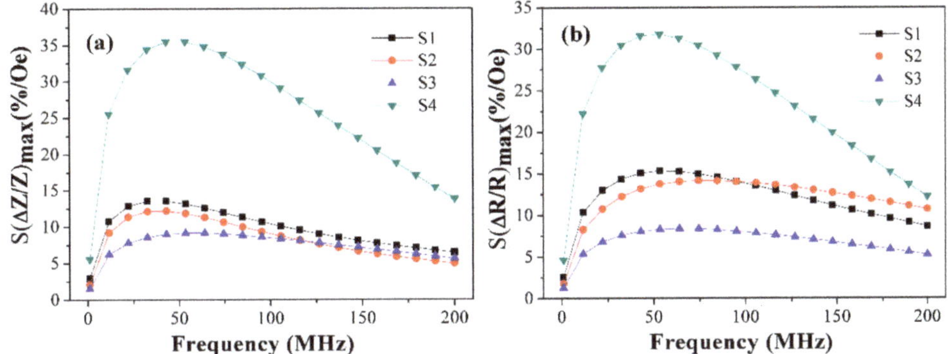

Figure 8. Frequency dependence of the sensitivity of the MI ratio for (**a**) total impedance and (**b**) its real part for $Co_{75}Fe_{10}Ni_2Si_8B_5$ meander samples.

In the frequency range 100-200 MHz, there is no obvious difference for $\Delta Z/Z_{max}$ ratios of samples S1 and S3. Compared with the six-turn meanders, the increase of length had a strong impact on MI enhancement. The maximum $\Delta R/R_{max}$ ratios was 254% obtained at a frequency of 53 MHz from the sample S4 (Figure 7b). The tendency for $(\Delta R/R)_{max}$ is similar to that of $(\Delta Z/Z)_{max}$. However, the $\Delta R/R_{max}$ ratios had a conspicuous distinction in the frequency range 1–200 MHz: the working frequencies correspond to the $\Delta R/R_{max}$ ratios for samples S1–S4 are higher than that $\Delta Z/Z_{max}$ ratios.

Figure 8 shows the frequency dependence of the sensitivity of the MI ratio for total impedance and its real part for different meanders. A general view of the frequency dependence of the MI maximum (both for real part and the total impedance) is quite similar to the $(\Delta R/R)_{max}$ $(\Delta Z/Z)_{max}$ measured for single stripe amorphous ribbons. The origin of such behavior was previously discussed in the literature. Under conditions of strong impedance variation, the constant external magnetic field and transverse magnetic field created by the alternate current make a contribution to the transverse dynamic permeability. With the frequency increase, the domain walls become affected by stronger eddy-current damping, the anisotropy field gradually increases, magnetic permeability decreases leading to decrease in MI value. More details can be found elsewhere [2,32–34].

The sensitivity for the total impedance MI ratio maxima $S(\Delta Z/Z)_{max}$ and for its real part maxima $S(\Delta R/R)_{max}$ were calculated in the field of 7 Oe, 14 Oe and 12 Oe respectively. The $\Delta Z/Z_{max}$ sensitivity of meander samples S4, S3, S2 and S1 were 36%/Oe, 9%/Oe 12%/Oe and 14%/Oe, respectively. These dependences reveal the same tendencies as those found for the frequency dependence the MI ratio maximum for the total impedance and its real part (Figure 8). The highest $S(\Delta Z/Z)_{max}$ and $S(\Delta R/R)_{max}$ (36%) were observed in the case of the sample S4 with increased frequency. The sensitivity of the MI ratio for total impedance and its real part of sample S3 were median. The values of sensitivity were also adequate for sensor applications from the point of view of magnetic field sensitivity.

Table 2 lists the sensitivity of the MI ratio for total impedance and its real part for different samples and the corresponding different magnetic fields and frequencies. The sample S4 had better MI characteristics including a wider range of the fields suitable for application. Therefore, the field and frequency dependence of the MI ratio of the S4 meander were investigated in more detail (Figure 9).

Measurements in the decreasing field after magnetic saturation were defined as "down" branch and the measurements in the increasing field after saturation were defined as the "up" branch.

Table 2. Comparison of sensitivity of the MI ratio for total impedance and its real part for $Co_{75}Fe_{10}Ni_2Si_8B_5$ meander samples, and the corresponding different magnetic fields and frequencies.

Sample	S ($\Delta Z/Z$)(%/Oe)	H (Oe)	f (MHz)	S ($\Delta R/R$)(%/Oe)	H (Oe)	f (MHz)
S1	14%	12	32	15	12	53
S2	12%	12	42	14	12	74
S3	9%	14	63	10	14	63
S4	36%	7	42	36	7	53

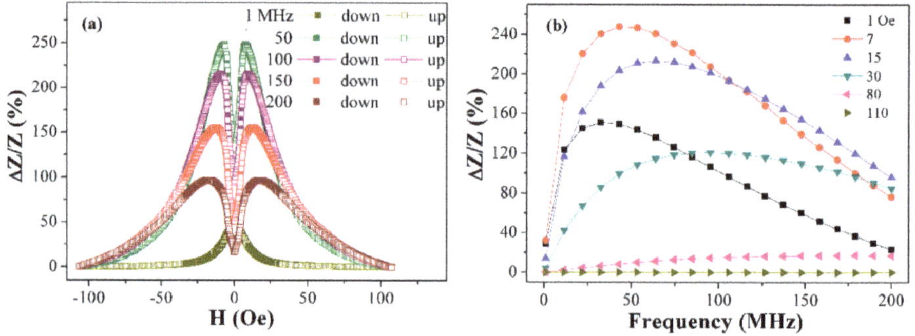

Figure 9. (a) Field and (b) frequency dependence of the MI ratio of $Co_{75}Fe_{10}Ni_2Si_8B_5$ patterned meander element S4.

The MI value is almost constant in higher fields due to the saturated sample state. Magnetoimpedance ratio in an external field of 30 Oe also varies slightly in the frequency range 40–200 MHz. A significant change in the MI ratio was observed by varying not only the magnitude of the external magnetic field from 0 to 15 Oe but also the frequency of the exciting current. The optimum mode of MI-detection by the S4 element is frequency range up to 40 MHz in low fields up to 15 Oe. As the biodetection of the magnetic markers requires the highest sensitivity with respect to the applied field, due to the very small value of the stray fields of magnetic nanoparticles employed as biomolecular labels [35], the S4 element seems to be the best candidate for this kind of application.

4. Conclusions

In summary, surface magnetic properties and the magnetoimpedance effect of patterned soft ferromagnetic $Co_{75}Fe_{10}Ni_2Si_8B_5$ ribbon-based meander sensor elements have been investigated. Samples with a different number of turns, line width and length were fabricated and studied. The results showed that the length of the samples plays the most important role in the formation of well-defined magnetic anisotropy, and hence a higher homogeneity and MI response. The maximum MI ratio of the total impedance of 250% for the exciting current frequency of 45 MHz, and its sensitivity to the external magnetic field of 36%/Oe were observed for the longest six-turn meander sample. These results can be used for further investigations and the production of magnetic field sensors based on the MI effect for the cases focused on large drop analysis.

Author Contributions: Z.Y. and G.V.K. conceived and designed the experiments; P.G. contributed to the fabrication and primary characterization of the sensor elements; S.O.V., Z.Y. and E.V.G. performed the MI measurements and their primary analysis; S.V.S. fabricated the sample holder and discussed the high frequency resultsand the manuscript; E.V.G., A.A.C. and Z.Y. performed the MOKE studies and analyzed the data; Z.Y., E.V.G., A.A.C., G.V.K. comparatively analyzed the whole set of the data, prepared graphical materials and wrote the manuscript and responses to Reviewers.

Funding: This research was funded in part by the grant Postdoc Ural Federal University UrFU-2018", in part by Natural Science Foundation of Henan Province of China (162300410233), in part they were obtained within the financial support for state assignment of Minobrnauki of Russia (theme "Alloys") and in part by µ4F ELKARTEK grant of The Basque Country Government.

Acknowledgments: The authors are grateful to the Laboratory of magnetic sensors of Ural Federal University and V.O. Vas'kovskiy for special support. We thank A.N. Gor'kovenko's for advanced training with MOKE measurements.

Conflicts of Interest: The authors declare no conflict of interest.

References

1. Landau, L.D.; Lifshitz, E.M. Course of theoretical physics. In *Electrodynamics of Continuous Media*; Elsevier Science: Amsterdam, The Netherlands, 2013; Volume 8.

2. Beach, R.; Berkowitz, A. Sensitive field-and frequency-dependent impedance spectra of amorphous FeCoSiB wire and ribbon. *J. Appl. Phys.* **1994**, *76*, 6209–6213. [CrossRef]

3. Panina, L.V.; Mohri, K.; Bushida, K.; Noda, M. Giant magneto-impedance and magneto-inductive effects in amorphous alloys. *J. Appl. Phys.* **1994**, *76*, 6198–6203. [CrossRef]

4. Makhotkin, V.E.; Shurukhin, B.P.; Lopatin, V.A.; Marchukov, P. Yu.; Levin, Yu. K. Magnetic field sensors based on amorphous ribbons. *Sens. Actuators A* **1991**, *21*, 759–762. [CrossRef]

5. Mohri, K.; Uchiyama, T.; Panina, L.V. Recent advances of micro magnetic sensors and sensing application. *Sens. Actuators A* **1997**, *59*, 1–8. [CrossRef]

6. Kurlyandskaya, G.V.; Portnov, D.S.; Beketov, I.V.; Larrañaga, A.; Safronov, A.P.; Orue, I.; Medvedev, A.I.; Chlenova, A.A.; Sanchez-Ilarduya, M.B.; Martinez-Amesti, A.; et al. Nanostructured materials for magnetic biosensing. *BBA Gen. Subj.* **2017**, *1861*, 1494–1506. [CrossRef]

7. Feng, Z.; Zhi, S.; Guo, L.; Wei, M.; Zhou, Y.; Lei, C. A novel integrated microfluidic platform based on micro-magnetic sensor for magnetic bead manipulation and detection. *Microfluid. Nanofluid.* **2018**, *22*, 86.

8. Buznikov, N.A.; Safronov, A.P.; Orue, I.; Golubeva, E.V.; Lepalovskij, V.N.; Svalov, A.V.; Chlenova, A.A.; Kurlyandskaya, G.V. Modelling of magnetoimpedance response of thin film sensitive element in the presence of ferrogel: Next step toward development of biosensor for in-tissue embedded magnetic nanoparticles detection. *Biosens. Bioelectron.* **2018**, *117*, 366–372. [CrossRef]

9. Machado, F.L.A.; de Argoldjo, A.E.P.; Puca, A.A.; Rodrigues, A.R.; Rezende, S.M. Surface magnetoimpedance measurements in soft-ferromagnetic materials. *Phys. Status Solidi A* **1999**, *173*, 135.

10. Kurlyandskaya, G.V.; Vazquez, M.; Munoz, J.L.; García, D.; McCord, J. Effect of induced magnetic anisotropy and domain structure features on magnetoimpedance in stress annealed Co-rich amorphous ribbons. *J. Magn. Magn. Mater.* **1999**, *196*, 259–261. [CrossRef]

11. Chen, L.; Bao, C.C.; Yang, H.; Li, D.; Lei, C.; Wang, T.; Hu, H.Y.; He, M.; Zhou, Y.; Cui, D.X. A prototype of giant magnetoimpedance-based biosensing system for targeted detection of gastric cancer cells. *Biosens. Bioelectron.* **2011**, *26*, 3246–3253. [CrossRef] [PubMed]

12. Ripka, P.; Zaveta, K. Magnetic sensors: Principles and applications. In *Handbook of Magnetic Materials*; Buschow, K.H.J., Ed.; Elsevier: Amsterdam, The Netherlands, 2009; Volume 18, pp. 347–420.

13. Kurlyandskaya, G.V.; Barandiaran, J.M.; Vazquez, M.; Garcia, D.; Dmitrieva, N.V. Infuence of geometrical parameters on the giant magnetoimpedance response in amorphous ribbons. *J. Magn. Magn. Mater.* **2000**, *215–216*, 740–742. [CrossRef]

14. Chaturvedi, A.; Dhakal, T.P.; Witanachchi, S.; Le, A.T.; Phan, M.H.; Srikanth, H. Critical length and giant magnetoimpedance in $Co_{69}Fe_{4.5}Ni_{1.5}Si_{10}B_{15}$ amorphous ribbons. *Mater. Sci. Eng. B* **2010**, *172*, 146–150. [CrossRef]

15. Yang, Z.; Wang, H.H.; Dong, X.W.; Yan, H.L.; Lei, C.; Luo, Y.S. Giant magnetoimpedance based immunoassay for cardiac biomarker myoglobin. *Anal. Methods* **2017**, *9*, 3636–3642. [CrossRef]

16. Rivero, M.A.; Maicas, M.; López, E.; Aroca, C.; Sánchez, M.C.; Sánchez, P. Influence of the sensor shape on permalloy/Cu/permalloy magnetoimpedance. *J. Magn. Magn. Mater.* **2003**, *254–255*, 636–638. [CrossRef]

17. Yuvchenko, A.A.; Lepalovskii, V.N.; Vas'kovskii, V.O.; Safronov, A.P.; Volchkov, S.O.; Kurlyandskaya, G.V. Magnetic impedance of structured film meanders in the presence of magnetic micro-and nanoparticles. *Tech. Phys.* **2014**, *59*, 230–236. [CrossRef]

18. Lodewijk, K.J.; Fernandez, E.; Garcia-Arribas, A.; Kurlyandskaya, G.V.; Lepalovskij, V.N.; Safronov, A.P.; Kooi, B.J. Magnetoimpedance of thin film meander with composite coating layer containing Ni nanoparticles. *J. Appl. Phys.* **2014**, *115*, 17A323. [CrossRef]
19. Antonov, A.S.; Gadetskii, S.N.; Granovskii, A.B.; D'yachkov, A.L.; Paramonov, V.P.; Perov, N.S.; Prokoshin, A.F.; Usov, N.A.; Lagar'kov, A.N. Giant magnetic impedance in amorphous and nanocrystalline multilayers. *Phys. Met. Metall.* **1997**, *83*, 61–71.
20. Fernández, E.; Lopez, A.; Garcia-Arribas, A.; Svalov, A.V.; Kurlyandskaya, G.V.; Barrainkua, A. High-frequency magnetoimpedance response of thin-film microstructures using coplanar waveguides. *IEEE Trans. Magn.* **2015**, *51*, 6100404. [CrossRef]
21. Vas'kovskii, V.O.; Savin, P.A.; Volchkov, S.O.; Lepalovskii, V.N.; Bukreev, D.A.; Buchkevich, A.A. Nanostructuring effects in soft magnetic films and film elements with magnetic impedance. *Tech. Phys.* **2013**, *58*, 105–110. [CrossRef]
22. Kolding, T.E. A four-step method for de-embedding gigahertz on-wafer CMOS measurements. *IEEE Trans. Electron Devices* **2000**, *47*, 734–740. [CrossRef]
23. Yang, Z.; Lei, J.; Lei, C.; Zhou, Y.; Wang, T. Effect of magnetic field annealing and size on the giant magnetoimpedance in micro-patterned Co-based ribbon with a meander structure. *Appl. Phys. A* **2014**, *116*, 1847–1851. [CrossRef]
24. Handrich, K.; Kobe, S. *Amorphe Ferro-and Ferri-Magnetika*; PhysikVerlag: Weinheim, Germany, 1980; Mir: Moscow, Rissia, 1982.
25. Mohri, K.; Uchiyama, T.; Panina, L.V.; Yamamoto, M.; Bushida, K. Recent advances of amorphous wire CMOS IC magneto-impedance sensors: Innovative high-performance micromagnetic sensor chip. *J. Sens.* **2015**, *2015*, 718069. [CrossRef]
26. Hernando, A.; Lopez-Dominguez, V.; Ricciardi, E.; Osiak, K.; Marin, P. Tuned scattering of electromagnetic waves by a finite length ferromagnetic microwire. *IEEE Trans. Antennas Propag.* **2016**, *64*, 1112–1115. [CrossRef]
27. Brzeska, M.; Panhorst, M.; Kamp, P.B.; Schotter, J.; Reiss, G.; Pühler, A.; Becker, A.; Brück, H. Detection and manipulation of biomolecules by magnetic carriers. *J. Biotechnol.* **2004**, *112*, 25–33. [CrossRef]
28. Amalou, F.; Gijs, M.A.M. Giant magnetoimpedance of chemically thinned and polished magnetic amorphous ribbons. *J. Appl. Phys.* **2001**, *90*, 3466–3470. [CrossRef]
29. Kurlyandskaya, G.V.; De Cos, D.; Volchkov, S.O. Magnetosensitive transducers for nondestructive testing operating on the basis of the giant magnetoimpedance effect: A review. *Russ. J. Nondestruct. Test.* **2009**, *45*, 377–398. [CrossRef]
30. Muñoz, J.L.; Kurlyandskaya, G.V.; Barandiarán, J.M.; Potapov, A.P.; Lukshina, V.A.; Vázquez, M. Nonuniform anisotropy and magnetoimpedance in stress annealed amorphous ribbons. *Phys. Met. Metall.* **2001**, *91*, S139–S142.
31. Pirota, K.R.; Kraus, L.; Knobel, M.; Pagliuso, P.G.; Rettori, C. Angular dependence of giant magnetoimpedance in an amorphous Co-Fe-Si-B ribbon. *Phys. Rev. B* **1999**, *60*, 6685. [CrossRef]
32. Yang, Z.; Lei, C.; Zhou, Y.; Sun, X.C. Study on the giant magnetoimpedance effect in micro-patterned Co-based amorphous ribbons with single strip structure and tortuous shape. *Microsyst. Technol.* **2015**, *21*, 1995–2001. [CrossRef]
33. Kurlyandskaya, G.V.; Barandiarán, J.M.; Muñoz, J.L.; Gutierrez, J.; Vazquez, M.; Garcia, D.; Vas'kovskiy, V.O. Frequency dependence of giant magnetoimpedance effect in CuBe/CoFeNi plated wire with different types of magnetic anisotropy. *J. Appl. Phys.* **2000**, *87*, 4822–4824. [CrossRef]
34. Kraus, L. The theoretical limits of giant magneto-impedance. *J. Magn. Magn. Mater.* **1999**, *196–197*, 354–356. [CrossRef]
35. Baselt, D.R.; Lee, G.U.; Natesan, M.; Metzger, S.W.; Sheehan, P.E.; Colton, R.J. A biosensor based on magnetoresistance technology. *Biosens. Bioelectron.* **1998**, *13*, 731–739. [CrossRef]

sensors

Article

Characterization and Relaxation Properties of a Series of Monodispersed Magnetic Nanoparticles

Yapeng Zhang [1], Jingjing Cheng [1],* and Wenzhong Liu [1,2,*

[1]   School of Artificial Intelligence and Automation, Huazhong University of Science and Technology, Wuhan 430074, China

[2]   Key Laboratory of Image Processing and Intelligent Control (Huazhong University of Science and Technology), Ministry of Education, Wuhan 430074, China

*   Correspondence: chengjj@hust.edu.cn (J.C.); lwz7410@hust.edu.cn (W.L.)

Received: 11 July 2019; Accepted: 31 July 2019; Published: 2 August 2019

Abstract: Magnetic iron oxide nanoparticles are relatively advanced nanomaterials, and are widely used in biology, physics and medicine, especially as contrast agents for magnetic resonance imaging. Characterization of the properties of magnetic nanoparticles plays an important role in the application of magnetic particles. As a contrast agent, the relaxation rate directly affects image enhancement. We characterized a series of monodispersed magnetic nanoparticles using different methods and measured their relaxation rates using a 0.47 T low-field Nuclear Magnetic Resonance instrument. Generally speaking, the properties of magnetic nanoparticles are closely related to their particle sizes; however, neither longitudinal relaxation rate r_1 nor transverse relaxation rate r_2 changes monotonously with the particle size d. Therefore, size can affect the magnetism of magnetic nanoparticles, but it is not the only factor. Then, we defined the relaxation rates r'_i (i = 1 or 2) using the induced magnetization of magnetic nanoparticles, and found that the correlation relationship between r'_1 relaxation rate and r_1 relaxation rate is slightly worse, with a correlation coefficient of R^2 = 0.8939, while the correlation relationship between r'_2 relaxation rate and r_2 relaxation rate is very obvious, with a correlation coefficient of R^2 = 0.9983. The main reason is that r_2 relaxation rate is related to the magnetic field inhomogeneity, produced by magnetic nanoparticles; however r_1 relaxation rate is mainly a result of the direct interaction of hydrogen nucleus in water molecules and the metal ions in magnetic nanoparticles to shorten the T_1 relaxation time, so it is not directly related to magnetic field inhomogeneity.

Keywords: magnetic nanoparticles; contrast agent; relaxation; relaxation rate; Langevin model; magnetic field inhomogeneity

---

1. Introduction

Magnetic iron oxide nanoparticles (MIONPs) have developed rapidly in recent years and have been widely used in biology, physics and medicine. They are quite small, usually nanoscale, and because of their scale, they can manifest many unique properties, such as superparamagnetism, i.e., when the applied magnetic field approaches zero, the induced magnetization and coercivity are also zero [1–4]. In addition, MIONPs have good temperature performance and can be used as temperature sensors. Some scholars used them to make some progress in the field of magnetic temperature measurement [5–11]. Due to their low toxicity, biocompatibility, and specificity after surface modification, MIONPs can be used as a target for drug delivery and disease treatment [12–15]. MIONPs are often used as contrast agents [16–20] in magnetic resonance imaging (MRI), which is one of the most important imaging methods in the field of medical diagnosis and scientific research [21–24]. MRI is a kind of imaging technology, which mainly uses the resonance effect of an electromagnetic field and hydrogen nucleus spin. Then, according to the collected magnetic resonance signal, imaging information from the tested object can be established. MRI not only has many imaging parameters and can be imaged in any direction without ionizing radiation damage to the human body, but can also perform non-invasive imaging of the internal structure or tissue of the human body or organism. According to the energy exchange between nuclear spin and the outside, hydrogen protons have two main relaxation mechanisms: longitudinal relaxation (spin-lattice relaxation) and transverse relaxation (spin-spin relaxation), which correspond to two relaxation time parameters, respectively: longitudinal relaxation time (T_1 relaxation time) and transverse relaxation time (T_2 relaxation time). These two relaxation mechanisms have a direct impact on the quality of magnetic resonance imaging. When the contrast agent (such as MIONPs) is added to the tested object, it will produce induced magnetization under the excitation of the magnetic field, which will affect the distribution and uniformity of the magnetic field around it, and then change the relaxation mechanisms of the protons around it; in short, the relaxation times will change. More intuitively, it will enhance the image contrast and speed up the image efficiency [23,25,26]. The enhancement effect of contrast agents on MRI can be expressed by the relaxation rate (r_1 relaxation rate and r_2 relaxation rate). Current research focuses on how to reduce proton T_1 relaxation time and T_2 relaxation time, how to improve the contrast performance of lesions and surrounding tissues, and how to accelerate the relaxation rate. In these studies, MIONPs can play a significant role. For example, different synthetic methods, size control, surface modification, etc. are used to improve their performance as a contrast agent. In addition, some scholars studied the use of different proportions of other metal materials such as Co, Zn and Mn doped iron oxide to enhance its saturation magnetization, thereby improving the relaxation enhancement properties of magnetic nanoparticles [27–30].

Characterization of the properties of magnetic nanoparticles plays an important role in the application of magnetic particles. In this paper, a series of commercial single core magnetic nanoparticles (SHP series, Ocean Nanotech, San Diego, CA, USA) with different nominal particle sizes (5 nm, 10 nm, 15 nm, 20 nm, 25 nm, 30 nm) were characterized using different methods. They can be stably monodispersed in aqueous solutions. Generally speaking, the properties of magnetic nanoparticles are closely related to their sizes, so we firstly measured the particle sizes using transmission electron microscopy and dynamic light scattering. Then the relaxation time of samples with different Fe ion concentrations was measured in a 0.47 T low-field nuclear magnetic resonance (LF-NMR) instrument and fitted to obtain r_1 relaxation rate and r_2 relaxation rate. However, neither r_1 relaxation rate calculated nor r_2 relaxation rate calculated changes monotonously with the particle size d. Because the addition of magnetic nanoparticles mainly changes the uniformity of the ambient magnetic field of the surrounding water molecules, and then changes the relaxation time, we further seek the relationship between the relaxation time and the induced magnetization of the magnetic nanoparticles in the magnetic field, and define the r_1' relaxation rate and r_2' relaxation rate. Because the T_2 relaxation process is very sensitive to the inhomogeneity of the magnetic field, and the addition of magnetic nanoparticles can directly affect the inhomogeneity of the magnetic field, it is found that the correlation between r_2' relaxation rate and r_2 relaxation rate is good. The effect of magnetic nanoparticle contrast agent on T_1 relaxation mechanism is mainly due to the direct interaction of the hydrogen nucleus in water molecules and metal ions in magnetic nanoparticles; therefore T_1 relaxation mechanism is not directly related to the magnetic field inhomogeneity. Therefore, the magnetic properties of magnetic nanoparticles are influenced by many factors, among which particle size is only one.

2. Materials and Methods

2.1. Magnetic Nanoparticles

We use an SHP series commercial magnetic nanoparticle reagent (Ocean NanoTech, San Diego, CA, USA), which is a single core magnetic nanoparticle. It is composed of ferric oxide magnetic nanoparticles, which are coated with carboxylic acid groups on the surface with good water solubility and dispersity. Its original iron ion concentration is 5 mg/mL. Six kinds of magnetic nanoparticle reagents were selected: SHP-05, SHP-10, SHP-15, SHP-20, SHP-25 and SHP-30 (nominal particle sizes are 5, 10, 15, 20, 25 and 30 nm, respectively, with a tolerance of < 2.5 nm). In addition, according to the measurement method of relaxation rate, samples with different Fe concentrations need to be prepared for each particle size of the magnetic nanoparticle reagent. Therefore, the magnetic nanoparticle reagent with the above particle size were diluted by deionized water to prepare five different Fe ion concentrations (1.79 mM, 1.12 mM, 0.89 mM, 0.45 mM, 0.22 mM), totaling 30 samples.

2.2. Transmission Electron Microscopy

The magnetic nanoparticles were imaged by transmission electron microscopy (TEM) (H-7000FA, HITACHI, Tokyo, Japan) with 110 kV to characterize their core size.

2.3. Dynamic Light Scattering

Dynamic Light Scattering (DLS) can detect the diffusion motion of particles and determine the hydrodynamic size (the overall size including magnetic core, polymer coating layer and the surrounding water layer) distribution of particles. Zetasizer Nano ZS90 (Malvern–Panalytical, Malvern, England) was used to measure the hydrodynamic size distribution of the above magnetic nanoparticles at a fixed scattering angle of 90 degree. The autocorrelation function of scattered light is analyzed, and the size distribution is calculated by assuming that the particle is an equivalent sphere.

2.4. LF-NMR

The relaxation time was measured using an LF-NMR instrument with a magnetic field of 0.47 T (MiniPQ001-20-15 mm, Niumag, Suzhou, China). The temperature of the main magnet and the sample chamber of the instrument is set at 35 °C To shorten the temperature balance time after the samples are put into the sample chamber, the samples are kept in an incubator set at 35 °C before the experiment. After starting the experiment, the samples were placed in the sample chamber of the LF-NMR instrument in turn for about 3 min, so that the temperature of the samples in the sample chamber could reach the thermal equilibrium as far as possible. The T_1 relaxation time is measured four times by using Inverse Recovery (IR) Sequence and then T_2 relaxation time was measured by using the CPMG (Carr-Purcell-Meiboom-Gill) sequence five consecutive times.

3. Results

3.1. Characterization of Magnetic Nanoparticles Samples: TEM, DLS

TEM: The TEM images of SHP-05, SHP-10, SHP-15, SHP-20, SHP-25 and SHP-30 magnetic nanoparticles samples in Figure 1 show that they are monodispersed magnetic nanoparticles.

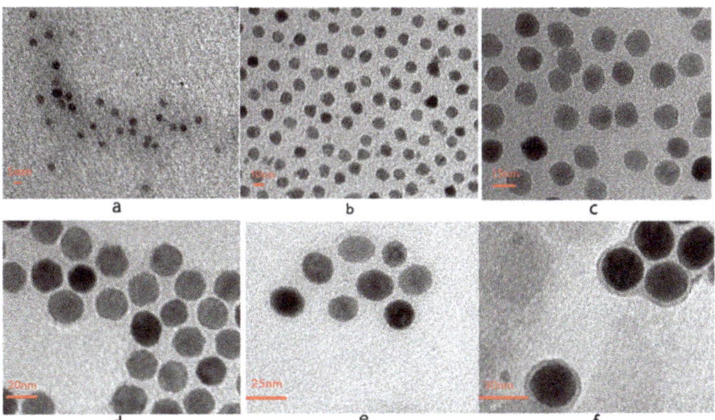

Figure 1. TEM images of SHP series magnetic nanoparticles samples. (**a**) SHP-05; (**b**) SHP-10; (**c**) SHP-15; (**d**) SHP-20; (**e**) SHP-25; (**f**) SHP-30.

DLS: The hydrodynamic size distributions are shown in Figure 2. To reduce the measurement error, each magnetic nanoparticle sample was measured three times in succession, and the statistical mean values were used.

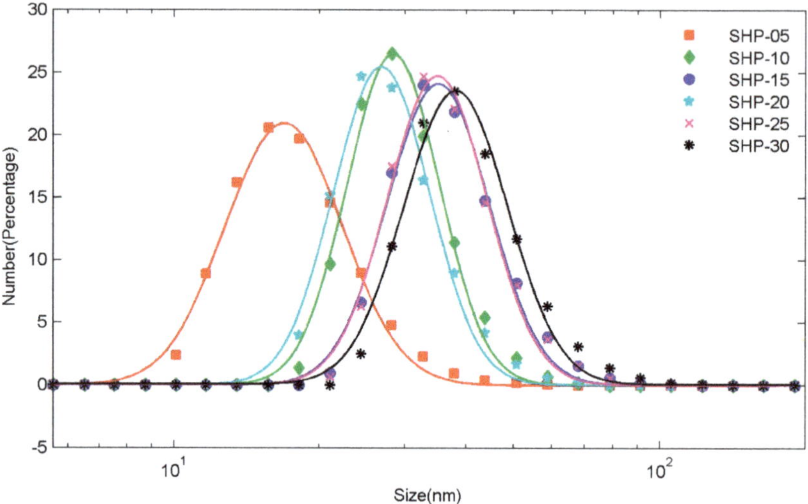

Figure 2. Hydrodynamic size distribution of SHP series magnetic nanoparticles. The discrete points are the measured hydrodynamic size distributions, and the solid lines are the fitting curve obtained using lognormal distribution.

Generally, we believe that the hydrodynamic size distribution of magnetic nanoparticles follows lognormal distribution [31–34]

$$f(d) = \frac{1}{d\sqrt{2\pi}\sigma} exp\left[-\frac{1}{2\sigma^2}\left(ln\frac{d}{\mu}\right)^2\right] \tag{1}$$

here, d is the diameter of particles, μ is the median diameter of the lognormal distribution, σ is the standard deviation of lnd. Therefore, the lognormal distribution function is used to fit the hydrodynamic size distribution measured by DLS, and the results are shown in Table 1.

Table 1. Parameters of lognormal distribution of the hydrodynamic size and the nominal size for SHP series magnetic nanoparticles.

Sample	DLS		Nominal Size (nm)
	Median Diameter (μ/nm)	**Variance (σ^2)**	
SHP-05	18.24	0.28	5
SHP-10	29.72	0.22	10
SHP-15	37.11	0.24	15
SHP-20	28.22	0.23	20
SHP-25	36.85	0.23	25
SHP-30	40.40	0.24	30

Dynamic light scattering (DLS) is a physical characterization method, which can be used to measure the particle size distribution of solutions or suspensions, and also to measure the behavior of complex fluids such as concentrated polymer solutions. The irregular random diffusion of magnetic particles in aqueous solution will attract some water molecules to move together on its surface, that is to say, water film is formed on its surface. It can be seen that the measurement results seem to be somewhat abnormal and have little correlation with the nominal core sizes. The results of DLS measurements using the same magnetic particles as us are not entirely consistent in other studies [35–37]. This may be caused by measurement errors, methods, etc., or more complex underlying causes.

3.2. Waiting Time Dependence of T_2 Relaxation Time

The relaxation time of each magnetic nanoparticle samples was then measured using a 0.47 T LF-NMR instrument. In addition, it was reported in much of the literature that the relaxation time (especially T_2 relaxation time) of magnetic nanoparticle aqueous solution samples obtained from NMR measurements is time-dependent [38–41]. That is to say, the measured relaxation times are related to the waiting time that the sample undergoes after being put into the sample chamber of the NMR instrument. In our experiment, because each sample was kept for about 3 min in the sample chamber, then the T_1 relaxation time was measured several times, and then the T_2 relaxation time was measured several times in succession. Therefore, depending on the above-mentioned articles on the waiting time dependence of T_2 relaxation time, this paper did not set the waiting time accurately. However, in order to illustrate the problem as much as possible, we define the waiting time t'_w, and set the waiting time t'_w for the first measurement of the T_2 relaxation time of each sample to 0 s. The waiting time t'_w of the same sample for subsequent T_2 relaxation time measurement was set according to the time interval from the first measurement. Generally, we choose the T_2 relaxation time measurement data of the sample with the highest Fe concentration for each particle size, and plot the relationship between T_2 relaxation time and waiting time t'_w, as shown in Figure 3 below.

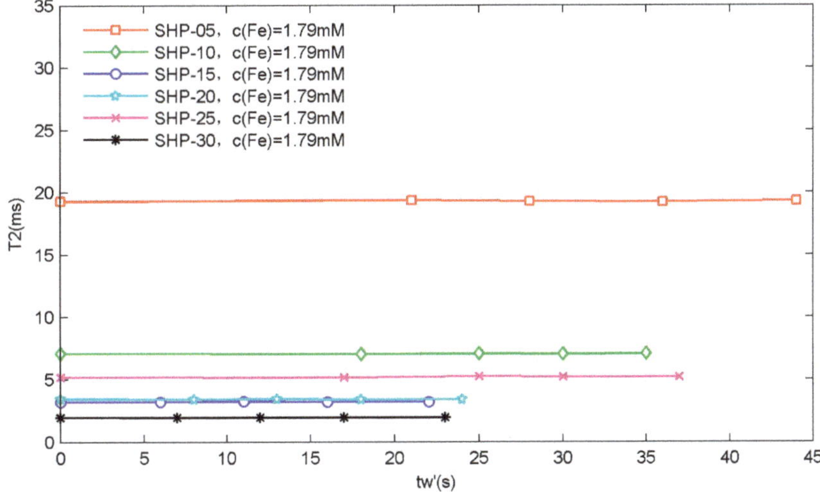

Figure 3. Waiting time dependence of T_2 relaxation time. It can be seen that the T_2 relaxation time of magnetic nanoparticle samples with different particle sizes hardly varies with the waiting time t'_w under the current test conditions.

It can be clearly seen that the T_2 relaxation time of the magnetic nanoparticles used in our experiment did not show dependence on t'_w waiting time. The main reason may be that the sizes of magnetic nanoparticle samples are relatively small, and surface-coated carboxylic acid groups produce electrostatic repulsion, which makes it possible for them to disperse stably in water phase without agglomeration under the effect of external magnetic field. Of course, it is also possible that the magnetic field (0.47 T) of the LF-NMR instrument in this paper is too low to make the magnetic nanoparticles in the magnetic field cluster into chains.

Regardless of the reason, the extrapolation method in these references is not needed in this paper [38–41], but the T_1 relaxation time and T_2 of relaxation time measured can be directly used to obtain the relaxation rate information of corresponding samples.

3.3. Relaxation Rate

In MRI, longitudinal relaxation rate r_1 and transverse relaxation rate r_2 are the main indicators to evaluate the enhancement effect of magnetic nanoparticles as contrast agent. Relaxation rate, in units of $mM^{-1}s^{-1}$, is the inverse of T_1 or T_2 relaxation time of protons when the concentration of Fe ion in magnetic nanoparticle dispersion solutions is 1 mM [39,42,43]:

$$1/T_{1,sup} = 1/T_{1,water} + r_1 c_{Fe} \tag{2}$$

$$1/T_{2,sup} = 1/T_{2,water} + r_2 c_{Fe} \tag{3}$$

where $T_{1,sup}$ and $T_{2,sup}$ are the T_1 and T_2 relaxation time of magnetic nanoparticle suspension solution respectively; $T_{1,water}$ and $T_{2,water}$ are the intrinsic T_1 and T_2 relaxation time of water respectively; c_{Fe}, in the unit of mM (mmol/L), is the concentration of Fe ion in magnetic nanoparticle suspension solution.

According to the above definition, we can simply prepare several magnetic nanoparticle water solution samples with different Fe ion concentrations, and use NMR instrument to measure their relaxation time respectively. Then taking $T_{i,water}$ (i = 1 or 2) and r_i (i = 1 or 2) as fitting parameters, by fitting the curve between the inverse of relaxation time and the concentration of Fe ion, the slope is the corresponding relaxation rate, as shown in Figure 4.

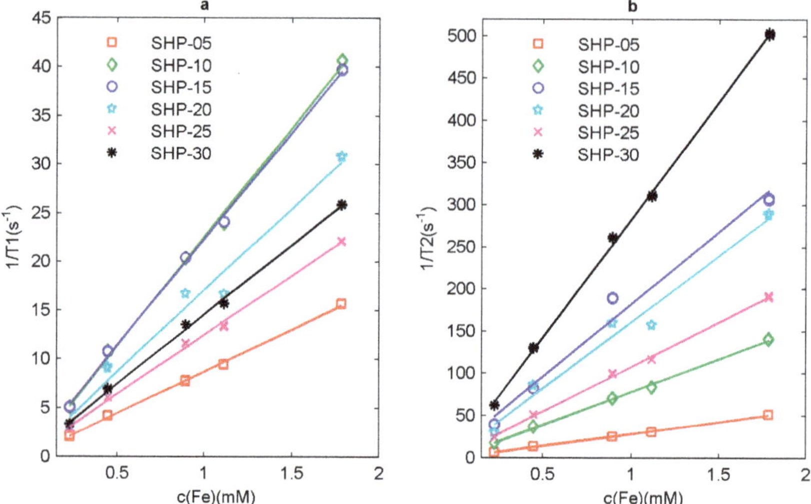

Figure 4. Relaxation rate of SHP series magnetic nanoparticle sample. (**a**) Inverse of longitudinal relaxation time $1/T_1$ and (**b**) inverse of transverse relaxation time $1/T_2$ with respect to Fe ion concentration c_{Fe}.

To make it more intuitive, we summarized the relaxation rate information of SHP series magnetic nanoparticle samples, as shown in Figure 5.

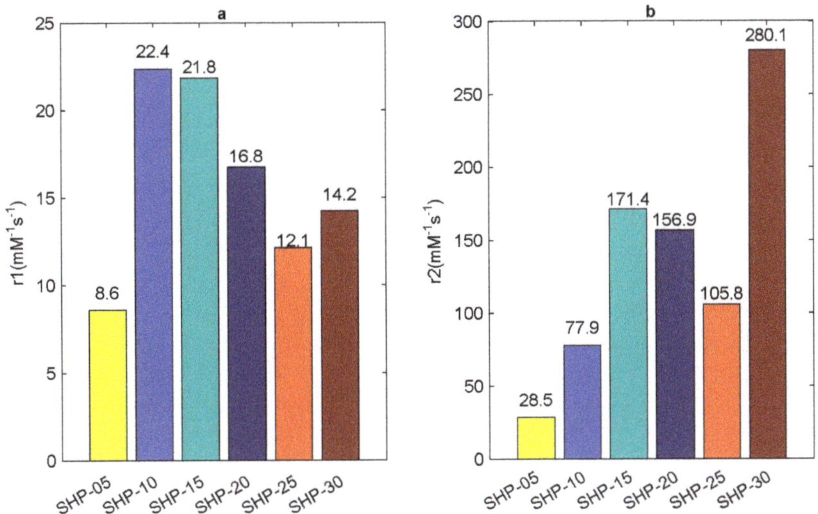

Figure 5. Relaxation rate of SHP series magnetic nanoparticle samples. (**a**) r_1 relaxation rate; (**b**) r_2 relaxation rate.

The addition of contrast agent will affect T_1 and T_2 relaxation time, but the degree of influence on the two relaxation times may be different, that is to say, r_1 relaxation rate and r_2 relaxation rate may be different, so we can simply classify the contrast agent using r_2 to r_1 ratio, i.e., r_2/r_1 [25,44–48]. It is generally believed that when the ratio $r_2/r_1 < 2$, the contrast agent works better as T_1 contrast agent; when $2 < r_2/r_1 < 10$, the contrast agent can work as both T_1 contrast agent and T_2 contrast agent, that is to say, it is T_1-T_2 dual mode contrast agent; when the ratio $r_2/r_1 > 10$, the contrast agent works better as T_2 contrast agent.

After calculating the r_2/r_1 ratio of the SHP series magnetic nanoparticles, and according to the above classification method, we then attempted to classify them, as shown in Figure 6.

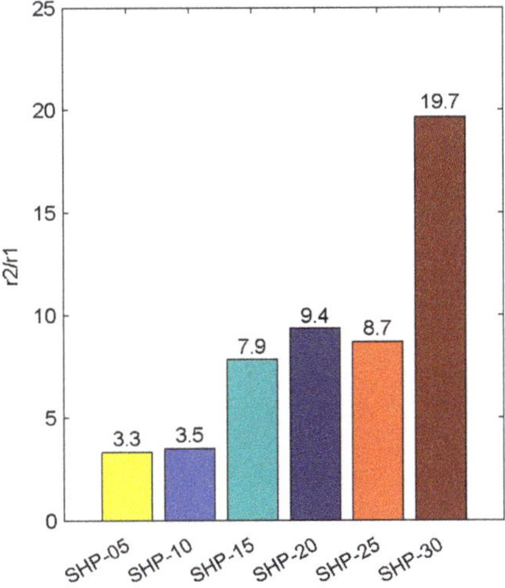

Figure 6. The ratio r_2/r_1 of SHP series magnetic nanoparticles.

It can be observed that the r_2/r_1 ratios of SHP-05, SHP-10, SHP-15, SHP-20 and SHP-25 magnetic nanoparticle samples are between 2 and 10, so they can be considered to be T_1-T_2 dual mode contrast agents. While the r_2/r_1 ratio of the SHP-30 magnetic nanoparticle sample is close to 20, it will work more suitably as T_2 contrast agent.

3.4. Analysis of Relaxation Rate

In addition, it can be found that neither r_1 relaxation rate calculated nor r_2 relaxation rate calculated changes monotonously with the particle size d [49,50]. When magnetic nanoparticles are added, the induced magnetization thereof will be produced under the exciting of the static magnetic field, which will change the ambient magnetic field, increase their inhomogeneity and shorten the relaxation time of the surrounding water hydrogen proton. This is also the basic principle of magnetic nanoparticles as contrast agents in MRI. It can be assumed that the magnetic field inhomogeneity is directly related to the induced magnetization produced by magnetic nanoparticles in magnetic field, and the induced magnetization can be described by the Langevin model [8,51–54]:

$$M = Nm\left(coth\left(\frac{mB}{kT}\right) - \frac{kT}{mB}\right) \qquad (4)$$

where N is the number of magnetic nanoparticles per unit volume (i.e., the concentration of magnetic nanoparticles); m is the magnetic moment of a single magnetic nanoparticle; B is the static magnetic field; k is the Boltzmann constant and T is the absolute temperature.

It can be seen that the induced magnetization of magnetic nanoparticles system in a static magnetic field is related to the external excitation magnetic field, temperature, concentration of magnetic particles and magnetic moment of magnetic particles. Generally, magnetic nanoparticles are simplified to sphere shapes, so the magnetic moment m of magnetic nanoparticle can be expressed as the following equation:

$$m = M_{sat}V = M_{sat}\frac{\pi d^3}{6} \qquad (5)$$

where, M_{sat} is the saturation magnetization of magnetic nanoparticle. d is the diameter of magnetic nanoparticles.

In the previous calculation of the relaxation rate, the concentration of Fe ion is taken as a reference. Here we assume that the magnetic field inhomogeneity caused by the addition of magnetic particles is proportional to the induced magnetization of magnetic nanoparticles, so we define the relaxation rate by referring to the induced magnetization of each sample. that is to say, we define r_i' (i = 1 or 2) relaxation rate. The method for calculating the r_i' (i = 1 or 2) relaxation rate is similar to that shown in Equations (2) and (3), except that the induced magnetization M (unit A.m) at different magnetic particle concentrations can be used to replace the Fe concentration c_{Fe}. When the induced magnetization intensity of MIONPs in water solution is 1 A.m, the inverse of T_1 relaxation time and T_2 relaxation time of the hydrogen proton is r_1' relaxation rate and r_2' relaxation rate respectively. It can be concluded that the unit of r_1' relaxation rate and r_2' relaxation rate is $A^{-1}m^{-1}s^{-1}$.

Therefore, according to the Langevin model, combined with the actual relaxation time measurement experiment, the induced magnetization of each magnetic nanoparticle sample in LF-NMR instrument is simulated. The static magnetic field B = 0.47 T (the main magnetic field of LF-NMR instrument MiniPQ001-20-15 mm), temperature T = 308.17 K (the temperature of the sample chamber of the LF-NMR instrument MiniPQ001-20-15 mm). In much of the literature, the saturation magnetization M_{sat} of SHP series magnetic nanoparticles was measured and involved, but the values measured (or assumed) are different, but the difference is not significant [35,55–58]. From Equation (5), it can be seen that magnetic moment m is proportional to the cube power of the particle size d, but is only proportional to the first power of saturation magnetization M_{sat}. In other words, the influence of magnetic particle size d on magnetic moment m is much bigger than the influence of the saturation magnetization M_{sat} on magnetic moment m. Therefore, we can say that the magnetic moment m is mainly dominated by particle size d. Therefore, according to the above references, it is reasonable to assume that the saturation magnetic moments of SHP series magnetic nanoparticles are all the same, i.e., $M_{sat} = 4.5 \times 10^5$ A/m.

In addition, the induced magnetization of magnetic particle system is also affected by the concentration of magnetic nanoparticles. Because the Langevin model needs concentration information from the magnetic nanoparticles, however the samples previously prepared are based on the concentration of Fe ion, so we need a conversion process. According to the Ocean Nanotech official website [59], the ratios of magnetic nanoparticle concentration to Fe ion concentration of SHP series magnetic nanoparticles, with different particle sizes, are different. After these conversion processes, finally, the induced magnetization M of different magnetic particle samples can be calculated using the Langevin model as shown in Equation (4). Then, by fitting the curve between the inverse of relaxation time and the induced magnetization M, the slope is the corresponding r_1' relaxation rate and r_2' relaxation rate, as shown in Figure 7. Moreover, the ratios of magnetic nanoparticle concentration to Fe ion concentration of every SHP reagent can be seen in Table A1 in Appendix A. In addition, parameters include magnetic nanoparticle concentration, induced magnetization, mean values and standard deviations of the measured relaxation time of the all tested magnetic nanoparticle samples are listed in Table A2 in Appendix A.

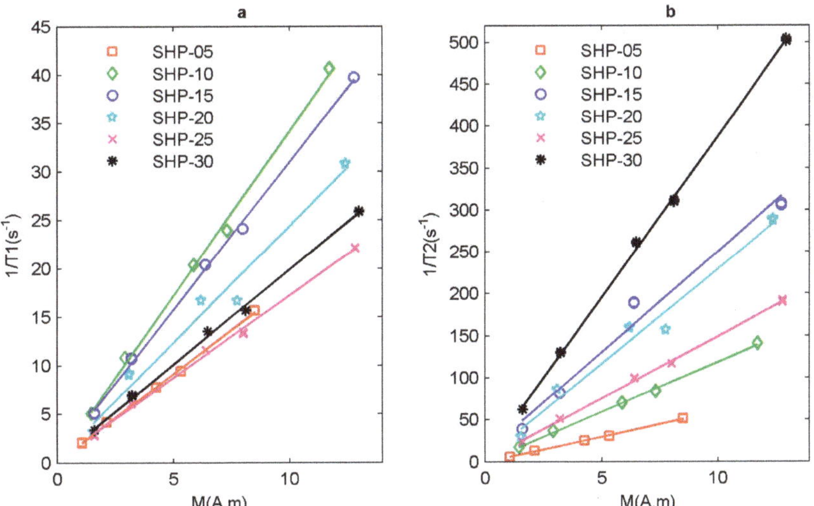

Figure 7. Relaxation rate of SHP series magnetic nanoparticle sample. (**a**) Inverse of longitudinal relaxation time $1/T_1$ and (**b**) inverse of transverse relaxation time $1/T_2$ with respect to the induced magnetization M.

Similarly, in order to be more intuitive, we summarized the r_1' relaxation rate and r_2' relaxation rate information of SHP series magnetic nanoparticle samples, as shown in Figure 8.

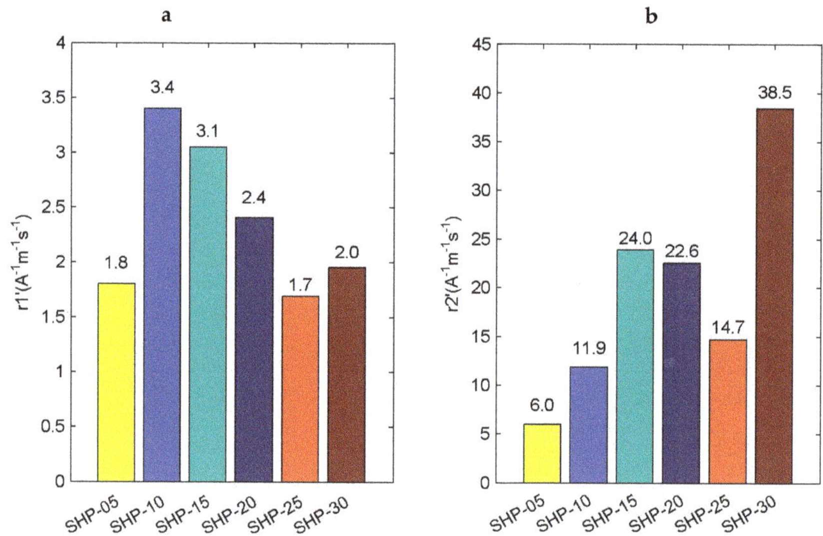

Figure 8. Relaxation rate of SHP series magnetic nanoparticle samples. (**a**) r_1' relaxation rate; (**b**) r_2' relaxation rate.

It can be seen that the r_1' relaxation rate or r_2' relaxation rate still have little relationship with the particle size. Next, we analyze the correlation between the r_i' (i = 1 or 2) relaxation rate and the r_i (i = 1 or 2) relaxation rate, and then linearly fit them, respectively, as shown in Figure 9. From the fitting results, it can be concluded that the correlation relationship between r_1' relaxation rate and r_1 relaxation rate is slightly worse, the correlation coefficient R^2 = 0.8939, while the correlation relationship between r_2' relaxation rate and r_2 relaxation rate is very obvious, the correlation coefficient R^2 = 0.9983.

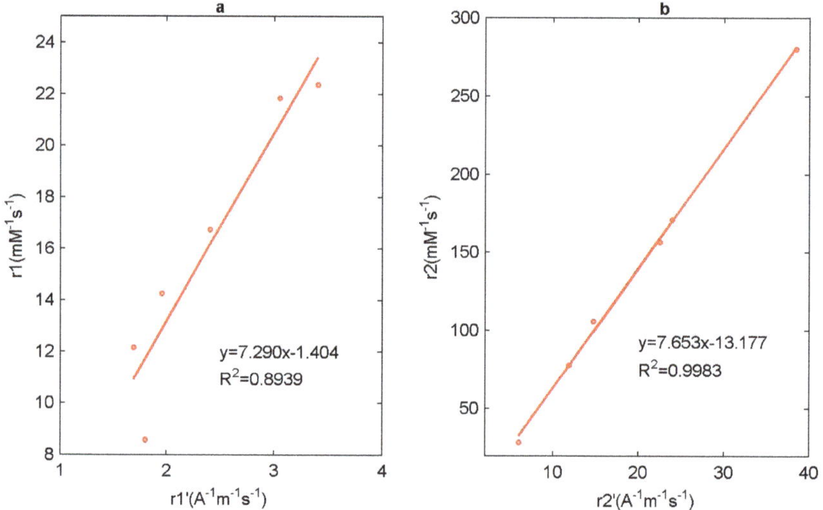

Figure 9. Linear regression of (**a**) r_1 relaxation rate with r_1' relaxation rate and (**b**) r_2 relaxation rate with r_2' relaxation rate.

The CPMG pulse sequence, designed for MRI and NMR, is designed to eliminate the influence of magnetic field inhomogeneity on the measurement of T_2 relaxation time as much as possible. Thus, we can obtain intrinsic T_2 relaxation time induced by spin-spin interaction. However, when magnetic nanoparticles are added to the aqueous solution sample, there are interactions between proton dipoles, proton dipoles and lattices, and between proton dipoles and electron dipoles of magnetic particles. These interactions will affect the relaxation mechanism of proton subsystems. In particular, magnetic nanoparticles will produce induced magnetization with the exciting of the magnetic field, which will change the magnetic field around them, aggravate the magnetic field inhomogeneity, accelerate the dephasing of hydrogen proton spin, and then change the relaxation of protons in the magnetic field. Moreover, the intensity of this impact will vary with the distance between hydrogen protons and magnetic particles. MIONPs with superparamagnetism can produce strong induced magnetization under the exciting magnetic field, which has a great influence on the inhomogeneity of the environmental magnetic field. Therefore, the T_2 relaxation time measured at this time includes all the above effects, and it can be said that it is mainly affected by the magnetic field inhomogeneity (the magnetic field inhomogeneity caused by the addition of magnetic nanoparticles) [60,61].

$$\frac{1}{T_{2,observe}} = \frac{1}{T_2} + \frac{1}{T_2'} = \frac{1}{T_2} + \gamma \Delta B \tag{6}$$

where $T_{2,observe}$ is the transverse relaxation time measured; T_2 is the intrinsic lateral relaxation time; T_2' is the transverse relaxation time due to the inhomogeneity of the magnetic field and can be expressed by $\gamma \Delta B$, where γ is the gyromagnetic ratio of the hydrogen proton, ΔB is the inhomogeneity of the magnetic field in the sample system and can be given by the Langevin model. From Equation (6) we can see that the relationship between $T_{2,observe}$ relaxation time and induced magnetization of magnetic nanoparticles system is obvious, so the correlation between r_2 relaxation rate and r_2' relaxation rate is good.

4. Discussion

As can be seen in Figure 5, basically, the smaller the particle size of magnetic nanoparticles, the larger the r_1 relaxation rate. The effect of magnetic nanoparticle contrast agent on T_1 relaxation mechanism is mainly due to the direct interaction of hydrogen nucleus in water molecules and the metal ions in magnetic nanoparticles to shorten the T_1 relaxation time, which can be explained using the Solomon-Bloembergen-Morgan theory (SBM) [62,63], so it is not directly related to magnetic field inhomogeneity. In short, it is related to the number of coordination water molecules, the exchange rate of water molecules, the rotation time of complexes and so on. As far as the metal ions (Fe) in magnetic nanoparticles are concerned, they have unpaired electrons, which interact with water electrons, such as by coordinating. The smaller the particle size, the larger the specific surface area S/V, and the easier the interaction between unpaired electrons and water electrons, so the larger the r_1 relaxation rate is. As far as r_2 relaxation rate is concerned, it is directly related to the magnetic field inhomogeneity caused by magnetic nanoparticles in this paper. The magnetization induced by magnetic nanoparticles in the magnetic field can be considered to be directly affecting the magnetic field inhomogeneity. The induced magnetization of magnetic nanoparticles is directly related to the magnetic moment ($m = M_{sat}\frac{\pi d^3}{6}$). Therefore, if the saturation magnetization M_{sat} of magnetic particles is constant, the larger the size of magnetic particles, the greater the induction magnetization and the larger the r_2 relaxation rate. Therefore, according to the respective influencing factors of r_1 relaxation rate and r_2 relaxation rate, it can be concluded that the larger the particle size, the larger the r_2/r_1, as shown in Figure 6. However, as far as the actual r_2 relaxation rate measured is concerned, it is basically shown that the larger the particle size is, the larger the r_2 relaxation rate is, but the data of SHP-20 and SHP-25 seem to be somewhat abnormal, as shown in Figure 5. The possible reason for this is that the saturation moments of SHP magnetic nanoparticles are not the same, and TEM imaging shows that the shape of SHP magnetic nanoparticles is not perfect spherical. Therefore, the magnetic properties of magnetic nanoparticles are affected by many factors, including composition, preparation process, surface coating, particle size, saturation magnetization and so on, while relaxation rate is only a simple characterization parameter. Moreover, Dynamic light scattering (DLS), which is a physical characterization method, and can be used to measure the particle size distribution of solutions or suspensions, and also to measure the behavior of complex fluids such as concentrated polymer solutions. The irregular random diffusion of magnetic particles in an aqueous solution will attract some water molecules to move together on its surface, that is to say, a water film is formed on its surface. The results of DLS measurements of magnetic nanoparticles used by us in different studies are not entirely consistent, which may be caused by their own measurement errors, methods and so on.

5. Conclusions

Magnetic nanoparticles are widely used as contrast agents for MRI. In this paper, SHP series magnetic nanoparticles (Ocean Nanotech) with different nominal sizes were selected for characterization experiments, and their relaxation rates were measured using a 0.47 T LF-NMR instrument. It was found that neither r_1 relaxation rate nor r_2 relaxation rate calculated changes monotonously with the particle size d. Size can affect the magnetic properties of magnetic nanoparticles, but it is not the only factor. There are other factors, such as morphology, agglomeration and so on. Furthermore, the induced magnetization of magnetic particle system in a static magnetic field can be calculated using the Langevin model, which serves as the source of magnetic field inhomogeneity in the measurement of relaxation time. Then the relationship between relaxation time and induced magnetization of magnetic nanoparticle samples in magnetic field is defined as r'_1 relaxation rate and r'_2 relaxation rate. The T_2 relaxation process is very sensitive to the magnetic field inhomogeneity, and the addition of magnetic nanoparticles can directly affect the magnetic field inhomogeneity. Therefore, it is found that the correlation between the r'_2 relaxation rate and the r_2 relaxation rate is very good, and the correlation coefficient reaches 0.9983. T_1 relaxation mechanism is not directly related to the magnetic field inhomogeneity. The effect of the magnetic nanoparticle contrast agent on T_1 relaxation mechanism is mainly due to the direct interaction of the hydrogen nucleus in water molecules and the metal ions in magnetic nanoparticles, so the correlation coefficient between the r'_2 relaxation rate and the r_2 relaxation rate is only 0.8939.

Author Contributions: Conceptualization, Y.Z., J.C. and W.L.; Data curation, Y.Z.; Formal analysis, Y.Z.; Methodology, Y.Z. and W.L.; Supervision, J.C. and W.L.; Writing—original draft, Y.Z.; Writing—review & editing, J.C. and W.L.

Funding: This study was supported by NSFC project 61571199.

Acknowledgments: We are grateful for the technical and equipment support of the NMR provided by Yiping Chen and Yongzhen Dong from Huazhong Agricultural University, Wuhan, China. In addition, we would like to thank Pei Zhang and Bi-Chao Xu from The Core Facility and Technical Support, Wuhan Institute of Virology, for their help with producing TEM micrographs.

Conflicts of Interest: The authors declare no conflict of interest.

Appendix A

We have sorted out the following two tables. Table A1 is the ratios of magnetic nanoparticle concentration to Fe ion concentration of every SHP reagent. Table A2 shows the parameters include magnetic nanoparticle concentration, induced magnetization, mean values and standard deviations of the measured relaxation time of the all tested magnetic nanoparticle samples.

Table A1. Fe concentrations and magnetic particle concentrations of the SHP series magnetic nanoparticles reagent. Data from the website of Ocean NanoTech.

Sample	Concentration of Fe (mg/mL)	Particle Amount (nmole) of 1 mg Fe
SHP-05	5	6.9
SHP-10	5	0.86
SHP-15	5	0.27
SHP-20	5	0.11
SHP-25	5	0.058
SHP-30	5	0.034

Table A2. Summary of information of all the magnetic nanoparticle samples in this paper.

MNP	No.	c_{Fe} (mM)	c_{MNP} (m^{-3})	Induced Magnetization M (A.m)	Mean of $1/T_2$ (s^{-1})	Standard Deviation of $1/T_2$ (s^{-1})	Mean of $1/T_1$ (s^{-1})	Standard Deviation of $1/T_1$ (s^{-1})
	1	1.79	4.154×10^{20}	8.51	51.80	0.10	15.68	0.03
	2	1.12	2.596×10^{20}	5.32	30.75	0.07	9.44	0.02
SHP-05	3	0.89	2.077×10^{20}	4.26	25.40	0.05	7.78	0.04
	4	0.45	1.038×10^{20}	2.13	13.65	0.03	4.20	0.01
	5	0.22	5.192×10^{19}	1.06	6.50	0.01	2.11	0.03
	6	1.79	5.177×10^{19}	11.73	141.10	0.37	40.67	0.04
	7	1.12	3.236×10^{19}	7.33	83.95	0.17	23.96	0.03
SHP-10	8	0.89	2.589×10^{19}	5.86	70.39	0.21	20.37	0.01
	9	0.45	1.294×10^{19}	2.93	36.86	0.09	10.83	0.00
	10	0.22	6.472×10^{18}	1.47	17.57	0.04	5.07	0.01
	11	1.79	1.625×10^{19}	12.78	307.03	0.85	39.73	0.03
	12	1.12	1.016×10^{19}	7.99			24.15	0.02
SHP-15	13	0.89	8.127×10^{18}	6.39	189.62	0.53	20.42	0.01
	14	0.45	4.064×10^{18}	3.19	82.40	0.19	10.78	0.05
	15	0.22	2.032×10^{18}	1.60	39.27	0.14	5.10	0.03
	16	1.79	6.622×10^{18}	12.42	289.04	0.79	30.85	0.03
	17	1.12	4.139×10^{18}	7.76	157.09	0.22	16.70	0.01
SHP-20	18	0.89	3.311×10^{18}	6.21	159.89	0.42	16.71	0.02
	19	0.45	1.656×10^{18}	3.11	86.34	0.22	9.16	0.07
	20	0.22	8.278×10^{17}	1.55	29.91	0.14	3.19	0.01
	21	1.79	3.492×10^{18}	12.82	191.39	0.88	22.16	0.03
	22	1.12	2.182×10^{18}	8.01	116.99	0.14	13.35	0.07
SHP-25	23	0.89	1.746×10^{18}	6.41	99.96	0.21	11.63	0.01
	24	0.45	8.729×10^{17}	3.21	51.62	0.23	6.09	0.01
	25	0.22	4.365×10^{17}	1.60	23.62	0.06	2.82	0.05
	26	1.79	2.047×10^{18}	13.00	502.91	1.21	25.90	0.03
	27	1.12	1.279×10^{18}	8.13	311.00	0.85	15.69	0.02
SHP-30	28	0.89	1.023×10^{18}	6.50	261.14	0.68	13.54	0.02
	29	0.45	5.117×10^{17}	3.25	130.26	0.40	6.93	0.05
	30	0.22	2.559×10^{17}	1.63	62.21	0.09	3.35	0.01

References

1. Klokkenburg, M.; Vonk, C.; Claesson, E.M.; Meeldijk, J.D.; Erné, B.H.; Philipse, A.P. Direct imaging of zero-field dipolar structures in colloidal dispersions of synthetic magnetite. *J. Am. Chem. Soc.* **2004**, *126*, 16706–16707. [CrossRef] [PubMed]

2. Kovalenko, M.V.; Bodnarchuk, M.I.; Lechner, R.T.; Günter, H.; Friedrich, S.F.; Wolfgang, H. Fatty acid salts as stabilizers in size- and shape-controlled nanocrystal synthesis: The case of inverse spinel iron oxide. *J. Am. Chem. Soc.* **2007**, *129*, 6352–6353. [CrossRef] [PubMed]

3. Jun, Y.-W.; Seo, J.-W.; Cheon, J. Nanoscaling laws of magnetic nanoparticles and their applicabilities in biomedical sciences. *Cheminform* **2008**, *39*, 179–189. [CrossRef]

4. Lang, X.Y.; Zheng, W.T.; Jiang, Q. Size and interface effects on ferromagnetic and antiferromagnetic transition temperatures. *Phys. Rev. B* **2006**, *73*, 224444. [CrossRef]

5. He, L.; Liu, W.; Xie, Q.; Pi, S.; Morais, P.C. A fast and remote magnetonanothermometry for a liquid environment. *Meas. Sci. Technol.* **2016**, *27*, 025901. [CrossRef]

6. Zhong, J.; Dieckhoff, J.; Schilling, M.; Ludwig, F. Influence of static magnetic field strength on the temperature resolution of a magnetic nanoparticle thermometer. *J. Appl. Phys.* **2016**, *120*, 143902. [CrossRef]

7. Weaver, J.B.; Rauwerdink, A.M.; Hansen, E.W. Magnetic nanoparticle temperature estimation. *Med. Phys.* **2009**, *36*, 1822. [CrossRef] [PubMed]

8. Du, Z.Z.; Su, R.J.; Liu, W.Z.; Huang, Z.X. Magnetic nanoparticle thermometer: An investigation of minimum error transmission path and AC bias error. *Sensors* **2015**, *15*, 8624–8641. [CrossRef] [PubMed]

9. Rauwerdink, A.M.; Hansen, E.W.; Weaver, J.B. Nanoparticle temperature estimation in combined ac and dc magnetic fields. *Phys. Med. Biol.* **2009**, *54*, L51. [CrossRef]

10. Zhong, J.; Liu, W.; Jiang, L.; Yang, M.; Morais, P.C. Real-time magnetic nanothermometry: The use of magnetization of magnetic nanoparticles assessed under low frequency triangle-wave magnetic fields. *Rev. Sci. Instrum.* **2014**, *85*, 1783–2887. [CrossRef]

11. Ming, Z.; Jing, Z.; Liu, W.; Du, Z.; Huang, Z.; Ming, Y.; Morais, P.C. Study of Magnetic Nanoparticle Spectrum for Magnetic Nanothermometry. *IEEE Trans. Magn.* **2015**, *51*, 1–6.

12. Bock, N.; Riminucci, A.; Dionigi, C.; Russo, A.; Tampieri, A.; Landi, E.; Goranov, V.A.; Marcacci, M.; Dediu, V. A novel route in bone tissue engineering: Magnetic biomimetic scaffolds. *Acta. Biomater.* **2010**, *6*, 786–796. [CrossRef] [PubMed]

13. Jiang, S.; Eltoukhy, A.A.; Love, K.T.; Langer, R.; Anderson, D.G. Lipidoid-Coated Iron Oxide Nanoparticles for Efficient DNA and siRNA delivery. *Nano. Lett.* **2013**, *13*, 1059–1064. [CrossRef] [PubMed]

14. Ito, A.; Shinkai, M.; Honda, H.; Kobayashi, T. Medical application of functionalized magnetic nanoparticles. *J. Biosci. Bioeng.* **2005**, *100*, 1–11. [CrossRef] [PubMed]

15. Sabale, S.; Kandesar, P.; Jadhav, V.; Komorek, R.; Motkuri, R.K.; Yu, X.Y. Recent developments in the synthesis, properties, and biomedical applications of core/shell superparamagnetic iron oxide nanoparticles with gold. *Biomater. Sci.* **2017**, *5*, 2212. [CrossRef] [PubMed]

16. Bu, L.; Xie, J.; Chen, K.; Huang, J.; Aguilar, Z.P.; Wang, A.; Sun, K.W.; Chua, M.S.; So, S.; Cheng, Z. Assessment and comparison of magnetic nanoparticles as MRI contrast agents in a rodent model of human hepatocellular carcinoma. *Contrast Media Mol. Imaging* **2012**, *7*, 363–372. [CrossRef] [PubMed]

17. Huang, J.; Wang, L.; Lin, R.; Wang, A.Y.; Yang, L.; Kuang, M.; Qian, W.; Mao, H. Casein-coated iron oxide nanoparticles for high MRI contrast enhancement and efficient cell targeting. *Appl. Mater. Interfaces* **2013**, *5*, 4632–4639. [CrossRef]

18. Dias, M.H.M.; Lauterbur, P.C. Ferromagnetic particles as contrast agents for magnetic resonance imaging of liver and spleen. *Magn. Reson. Med.* **2010**, *3*, 328–330. [CrossRef]

19. Werner, E.J.; Datta, A.; Jocher, C.J.; Raymond, K.N. High-relaxivity MRI contrast agents: Where coordination chemistry meets medical imaging. *Angew. Chem.* **2008**, *47*, 8568–8580. [CrossRef]

20. Xie, J.; Huang, J.; Li, X.; Sun, S.; Chen, X. Iron oxide nanoparticle platform for biomedical applications. *Curr. Med. Chem.* **2009**, *16*, 1278–1294. [CrossRef]

21. Henson, R. Introduction to functional magnetic resonance imaging: Principles and techniques. *Acta Radiol.* **2002**, *43*, 2110. [CrossRef]

22. Brown, M.A.; Semelka, R.C. *MRI: Basic Principles and Applications*, 2rd ed.; John Wiley & Sons: Hoboken, NJ, USA, 2015.

23. Mody, V.V.; Nounou, M.I.; Bikram, M. Novel nanomedicine-based MRI contrast agents for gynecological malignancies. *Adv. Drug Deliv. Rev.* **2009**, *61*, 795–807. [CrossRef] [PubMed]

24. Callaghan, P.T. *Principles of Nuclear Magnetic Resonance Microscopy*; Oxford University Press: New York, NY, USA, 1991.

25. Caravan, P.; Ellison, J.J.; Mcmurry, T.J.; Lauffer, R.B. Gadolinium(III) Chelates as MRI Contrast Agents: Structure, Dynamics, and Applications. *Cheminform* **2010**, *30*, 2293–2352. [CrossRef]

26. Lauffer, R.B. Paramagnetic metal complexes as water proton relaxation agents for NMR imaging: Theory and design. *Chem.. Rev.* **1987**, *87*, 901–927. [CrossRef]

27. Lee, J.-H.; Huh, Y.-M.; Jun, Y.-W.; Seo, J.-W.; Jang, J.-T.; Song, H.-T.; Kim, S.-J.; Cho, E.-J.; Yoon, H.-G.; Suh, J.-S.; et al. Artificially engineered magnetic nanoparticles for ultra-sensitive molecular imaging. *Nat. Med.* **2007**, *13*, 95–99. [CrossRef] [PubMed]

28. Jang, J.-T.; Nah, H.; Lee, J.-H.; Moon, S.H.; Kim, M.G.; Cheon, J. Critical enhancements of MRI contrast and hyperthermic effects by dopant-controlled magnetic nanoparticles. *Angew. Chem.* **2009**, *121*, 1260–1264. [CrossRef]

29. Miclea, C.; Tanasoiu, C.; Miclea, C.F.; Spanulescu, I.; Cioangher, M.; Miclea, C.T. Magnetic Temperature Transducers Made from Copper Based Soft Ferrite. *Adv. Sci. Technol.* **2008**, *54*, 62–69. [CrossRef]

30. Hankiewicz, J.H.; Alghamdi, N.; Hammelev, N.M.; Anderson, N.R.; Camley, R.E.; Stupic, K.; Przybylski, M.; Zukrowski, J.; Celinski, Z.J. Zinc doped copper ferrite particles as temperature sensors for magnetic resonance imaging. *Aip. Adv.* **2017**, *7*, 56703. [CrossRef]

31. Sherman, S.G.; Wereley, N.M. Effect of Particle Size Distribution on Chain Structures in Magnetorheological Fluids. *IEEE Trans. Magn.* **2013**, *49*, 3430–3433. [CrossRef]

32. Liu, W.Z.; Zhou, M.; Li, K. Estimation of the size distribution of magnetic nanoparticles using modified magnetization curves. *Meas. Sci. Technol.* **2009**, *20*, 125802. [CrossRef]

33. Ferguson, R.M.; Minard, K.R.; Khandhar, A.P.; Krishnan, K.M. Optimizing magnetite nanoparticles for mass sensitivity in magnetic particle imaging. *Med. Phys.* **2011**, *38*, 1619–1626. [CrossRef] [PubMed]

34. Lak, A.; Ludwig, F.; Scholtyssek, J.M.; Dieckhoff, J.; Fiege, K.; Schilling, M. Size Distribution and Magnetization Optimization of Single-Core Iron Oxide Nanoparticles by Exploiting Design of Experiment Methodology. *IEEE Trans. Magn.* **2013**, *49*, 201–207. [CrossRef]

35. Adolphi, N.L.; Huber, D.L.; Bryant, H.C.; Monson, T.C.; Fegan, D.L.; Jitkang, L.; Trujillo, J.E.; Tessier, T.E.; Lovato, D.M.; Butler, K.S. Characterization of single-core magnetite nanoparticles for magnetic imaging by SQUID relaxometry. *Phys. Med. Biol.* **2012**, *55*, 5985–6003. [CrossRef] [PubMed]

36. Østerberg, F.W.; Rizzi, G.; Hansen, M.F. On-chip measurements of Brownian relaxation of magnetic beads with diameters from 10 nm to 250 nm. *J. Appl. Phys.* **2013**, *113*, 8130–8183. [CrossRef]

37. Tong, H.; Kang, W.; Shi, Y.; Zhou, G.; Lu, Y. Physiological function and inflamed-brain migration of mouse monocyte-derived macrophages following cellular uptake of superparamagnetic iron oxide nanoparticles—Implication of macrophage-based drug delivery into the central nervous system. *Int. J. Pharm.* **2016**, *505*, 271–282. [CrossRef] [PubMed]

38. Chen, D.X.; Sun, N.; Huang, Z.J.; Cheng, C.M.; Xu, H.; Gu, H.C. Experimental study on T2 relaxation time of protons in water suspensions of iron-oxide nanoparticles: Effects of polymer coating thickness and over-low. *J. Magn. Magn. Mater.* **2010**, *322*, 548–556. [CrossRef]

39. Sun, N.; Chen, D.X.; Gu, H.C.; Wang, X.L. Experimental study on T2 relaxation time of protons in water suspensions of iron-oxide nanoparticles: Waiting time dependence. *J. Magn. Magn. Mater.* **2009**, *321*, 2971–2975. [CrossRef]

40. Xu, F.; Gu, H. Experimental study on transverse relaxation rate of protons in aqueous suspensions of magnetite nanocrystal clusters with a SiO_2 shell. *J. Magn. Magn. Mater.* **2013**, *343*, 60–64. [CrossRef]

41. Chen, D.X.; Xu, F.J.; Gu, H.C. Experimental study on transverse relaxation rate of protons in water suspensions of magnetite nanoclusters: Dependence of cluster sizes, volume fraction, inter-echo time, and waiting time. *J. Magn. Magn. Mater.* **2012**, *324*, 2809–2820. [CrossRef]

42. Roch, A.; Bach-Gansmo, T.; Muller, R.N. In vitro relaxometric characterization of superparamagnetic contrast agents. *Magma. Magn. Reson. Mater. Phys. Biol. Med.* **1993**, *1*, 83–88. [CrossRef]

43. Stark, D.D.; Weissleder, R.; Elizondo, G.; Hahn, P.F.; Saini, S.; Todd, L.E.; Wittenberg, J.; Ferrucci, J.T. Superparamagnetic iron oxide: Clinical application as a contrast agent for MR imaging of the liver. *Radiology* **1988**, *168*, 297–301. [CrossRef]

44. Lee, N.; Hyeon, T. Designed synthesis of uniformly sized iron oxide nanoparticles for efficient magnetic resonance imaging contrast agents. *Cheminform* **2012**, *41*, 2575–2589.

45. Hu, F.Q.; Jia, Q.J.; Li, Y.L.; Gao, M.Y. Facile synthesis of ultrasmall PEGylated iron oxide nanoparticles for dual-contrast T1- and T2-weighted magnetic resonance imaging. *Nanotechnology* **2011**, *22*, 245604. [CrossRef]

46. Lee, N.; Yoo, D.; Ling, D.; Cho, M.H.; Hyeon, T.; Cheon, J. Iron Oxide Based Nanoparticles for Multimodal Imaging and Magnetoresponsive Therapy. *Chem. Rev.* **2015**, *115*, 10637. [CrossRef]

47. Kim, B.; Lee, N.; Kim, H.; An, K.; Park, Y.; Choi, Y.; Shin, K.; Lee, Y.; Kwon, S.G.; Na, H.B.; et al. Large-scale synthesis of uniform and extremely small-sized iron oxide nanoparticles for high-resolution T1 magnetic resonance imaging contrast agents. *J. Am. Chem. Soc.* **2011**, *133*, 12624–12631. [CrossRef]

48. Leal, M.P.; Rivera-Fernández, S.; Franco, J.M.; Pozo, D.; de la Fuente, J.M.; García-Martín, M.L. Long-circulating PEGylated manganese ferrite nanoparticles for MRI-based molecular imaging. *Nanoscale* **2014**, *7*, 2050–2059. [CrossRef]

49. Roch, A.; Gossuin, Y.; Muller, R.N.; Gillis, P. Superparamagnetic colloid suspensions: Water magnetic relaxation and clustering. *J. Magn. Magn. Mater.* **2005**, *293*, 532–539. [CrossRef]

50. Huang, G.M.; Li, H.; Chen, J.H.; Zhao, Z.H.; Yang, L.J.; Chi, X.Q.; Chen, Z.; Wang, X.M.; Gao, J.H. Tunable T1 and T2 contrast abilities of manganese-engineered iron oxide nanoparticles through size control. *Nanoscale* **2014**, *6*, 10404–10412. [CrossRef]

51. Pi, S.; Liu, W.; Wei, K.; Mosiniewicz-Szablewska, E. AC Magnetic Nanothermometry: An Investigation of the Influence of Size Distribution of Magnetic Nanoparticles. *IEEE Trans. Magn.* **2017**, *53*, 1–7. [CrossRef]

52. Enpuku, K.; Tsujita, Y.; Nakamura, K.; Sasayama, T.; Yoshida, T. Biosensing utilizing magnetic markers and superconducting quantum interference devices. *Supercond. Sci. Technol.* **2017**, *30*, 53002. [CrossRef]

53. Yoshida, T.; Ogawa, K.; Tsubaki, T.; Othman, N.B.; Enpuku, K. Detection of Magnetic Nanoparticles Using the Second-Harmonic Signal. *IEEE Trans. Magn.* **2011**, *47*, 2863–2866. [CrossRef]

54. Stamps, R.L. *Magnetism of Surfaces, Interfaces, and Nanoscale Materials*; Elsevier Science: Amsterdam, Holland, 2016.

55. Hathaway, H.J.; Butler, K.S.; Adolphi, N.L.; Lovato, D.M.; Belfon, R.; Fegan, D.; Monson, T.C.; Trujillo, J.E.; Tessier, T.E.; Bryant, H.C. Detection of breast cancer cells using targeted magnetic nanoparticles and ultra-sensitive magnetic field sensors. *Breast Cancer Res.* **2011**, *13*, R108. [CrossRef]

56. Draack, S.; Viereck, T.; Nording, F.; Janssen, K.J.; Schilling, M.; Ludwig, F. Determination of dominating relaxation mechanisms from temperature-dependent Magnetic Particle Spectroscopy measurements. *J. Magn. Magn. Mater.* **2019**, *474*, 570–573. [CrossRef]

57. Ahrentorp, F.; Astalan, A.; Blomgren, J.; Jonasson, C.; Wetterskog, E.; Svedlindh, P.; Lak, A.; Ludwig, F.; Ijzendoorn, L.J.V.; Westphal, F. Effective particle magnetic moment of multi-core particles. *J. Magn. Magn. Mater.* **2015**, *380*, 221–226. [CrossRef]

58. Ludwig, F.; Kazakova, O.; Barquin, L.F.; Fornara, A.; Johansson, C. Magnetic, Structural, and Particle Size Analysis of Single- and Multi-Core Magnetic Nanoparticles. *IEEE Trans. Magn.* **2014**, *50*, 1–4. [CrossRef]

59. Ocean NanoTech. Conjugation Protocol of Carboxylic Acid Functionalized Iron Oxide Nanoparticles (SHP). Available online: https://www.oceannanotech.com/media/wysiwyg/protocol/SHP_5-30_PTC.pdf (accessed on 24 June 2019).

60. Chavhan, G.B.; Babyn, P.S.; Thomas, B.; Shroff, M.M.; Haacke, E.M. Principles, techniques, and applications of T2*-based MR imaging and its special applications. *Radiographics* **2009**, *29*, 1433–1449. [CrossRef]

61. Na, H.B.; Song, I.C.; Hyeon, T. Inorganic Nanoparticles for MRI Contrast Agents. *Adv. Mater.* **2009**, *21*, 2133–2148. [CrossRef]

62. Zhou, Z.; Zhao, Z.; Zhang, H.; Wang, Z.; Chen, X.; Wang, R.; Chen, Z.; Gao, J. Interplay between Longitudinal and Transverse Contrasts in Fe_3O_4 Nanoplates with (111) Exposed Surfaces. *ACS Nano* **2014**, *8*, 7976. [CrossRef]

63. Manus, L.M.; Strauch, R.C.; Hung, A.H.; Eckermann, A.L.; Meade, T.J. Analytical Methods for Characterizing Magnetic Resonance Probes. *Anal. Chem.* **2012**, *84*, 6278–6287. [CrossRef]

Article

Ferrogels Ultrasonography for Biomedical Applications

Felix A. Blyakhman [1,2], Sergey Yu Sokolov [1,2], Alexander P. Safronov [2,3], Olga A. Dinislamova [1], Tatyana F. Shklyar [1,2], Andrey Yu Zubarev [2,4] and Galina V. Kurlyandskaya [2,5,*]

[1]   Ural State Medical University, 620028 Ekaterinburg, Russia; feliks.blyakhman@urfu.ru (F.A.B.); sergey.sokolov@urfu.ru (S.Y.S.); o_dinislamova@rambler.ru (O.A.D.); t.f.shkliar@urfu.ru (T.F.S.)

[2]   Institute of Natural Sciences and Mathematics Ural Federal University, 620002 Ekaterinburg, Russia; safronov@iep.uran.ru (A.P.S.); A.J.Zubarev@urfu.ru (A.Y.Z.)

[3]   Institute of Electrophysics, Ural Division RAS, 620016 Ekaterinburg, Russia

[4]   M.N. Mikheev Institute of Metal Physics of the Ural Branch of the Russian Academy of Sciences, 620990 Ekaterinburg, Russia

[5]   Departamento de Electricidad y Electrónica and BCMaterials, Universidad del País Vasco UPV/EHU, 48080 Bilbao, Spain

*   Correspondence: galina@we.lc.ehu.es; Tel.: +34-9460-13237; Fax: +34-9460-13071

Received: 16 July 2019; Accepted: 10 September 2019; Published: 13 September 2019

Abstract: Ferrogels (FG) are magnetic composites that are widely used in the area of biomedical engineering and biosensing. In this work, ferrogels with different concentrations of magnetic nanoparticles (MNPs) were synthesized by the radical polymerization of acrylamide in stabilized aqueous ferrofluid. FG samples were prepared in various shapes that are suitable for different characterization techniques. Thin cylindrical samples were used to simulate the case of targeted drug delivery test through blood vessels. Samples of larger size that were in the shape of cylindrical plates were used for the evaluation of the FG applicability as substitutes for damaged structures, such as bone or cartilage tissues. Regardless of the shape of the samples and the conditions of their location, the boundaries of FG were confidently visualized over the entire range of concentrations of MNPs while using medical ultrasound. The amplitude of the reflected echo signal was higher for the higher concentration of MNPs in the gel. This result was not related to the influence of the MNPs on the intensity of the reflected echo signal directly, since the wavelength of the ultrasonic effect used is much larger than the particle size. Qualitative theoretical model for the understanding of the experimental results was proposed while taking into account the concept that at the acoustic oscillations of the hydrogel, the macromolecular net, and water in the gel porous structure experience the viscous Stocks-like interaction.

Keywords: magnetic nanoparticles; ferrogels; medical ultrasound; sonography; biomedical applications

---

1. Introduction

Hydrogels are soft materials that are widely used in the area of biomedical applications [1]. Ferrogels (FG) are composites that contain a polymer swollen in a solvent and filled with nano- or micro-sized magnetic particles [2–4]. In particular, FG based on polyacrylamide (PAA) hydrogel with magnetic nanoparticles (MNPs) of iron oxides are the most studied and sought-after material for a wide range of biomedical applications, including magnetic biosensors, drug delivery, and regenerative medicine [5,6]. Different studies had established the fact that the elastic properties of PAA ferrogels can significantly vary, depending on the details of synthesis conditions, as well as change their mechanical characteristics in response to an external magnetic field [4,7,8]. It is also known that polyacrylamide ferrogels have low toxicity and good compatibility to living cells [9,10]. Based on this advantage, PAA ferrogels have been used as substrates (scaffolds) for cell culturing for the needs of cellular technologies and tissue engineering [11].

Ferrogel scaffolds offer different directions of the research and applications. For example, the intensity of ultrasound that is reflected at a ferrogel/water interface is determined by the difference between the acoustic impedances of the two materials. The effect of magnetic nanoparticles on the ultrasonic parameters (also acoustic impedance) of gel phantoms has been previously shown [12]. The values of the impedance increased after the addition of nanoparticles, so the phantoms with magnetic nanoparticles exhibited increased echogenicity, owing to the significant number of scatters [13].

In addition, the MNPs forming the part of the PAA gels have high magnetic responses, often sufficient for the control of the movement of FG based micro-objects while using an external magnetic field [14]. This feature opens up prospects for the use of ferrogels for magnetic biosensor applications, regenerative medicine, or targeted drug delivery or controlled release of drugs [15–17]. The above-mentioned applications are associated with the solution of a number of problems, including the task of reliable visualization of FG in a living organism.

In fact, the scientific community has already moved from the concept of biomedical applications of magnetic nanoparticles toward the understanding of ferrogel models, which are much closer to realistic applications in which MNPs are distributed in living tissue. Very often, such distribution reflects the biological properties of the natural tissue. What is especially important, aggregation features can be conditioned by irregularly structured tissue with the disease-affected morphology [18]. The development of a new technique for detecting of ferrogel scaffolds in blood vessels while using medical ultrasound is highly desired. There is a strong request for FG use in vivo as a prototype of magnetically controlled platforms for targeted delivery of cell implants and drug substrates through arteries, as well as positioning indicators for low invasive surgery.

In the work, the features of ultrasonic location of samples of PAA based ferrogels with a variation of the concentration of Fe_2O_3 MNPs of iron oxide in a wide range are considered for FG scaffold of different shape. A theoretical interpretation of the experimental results is given.

2. Experimental

2.1. Gel and Ferrogel Synthesis and Characterization

Ferrogels with varying concentration of iron oxide magnetic nanoparticles were synthesized by three-dimensional (3D) radical polymerization of acrylamide in stabilized aqueous ferrofluid. First, electrostatically stabilized ferrofluid was prepared with stock concentration 5.1% of MNPs by weight. Fe_2O_3 MNPs (Figure 1a) were synthesized by the laser target evaporation (LTE) method while using commercial magnetite (Fe_3O_4) (Alfa Aesar, Ward Hill, MA, USA) as a precursor.

Technological details of the LTE method were previously described in earlier studies [19,20].

Stock ferrofluid for the synthesis of FGs was prepared by the ultrasound dispersion of maghemite MNPs in 5 mM solution of sodium citrate, taken as electrostatic dispersant. Ferrofluid was centrifuged for 5 min. at 10,000 rpm to eliminate large aggregates. The number averaged diameter of particles in ferrofluid determined by dynamic light scattering (Brookhaven ZetaPlus, Brookhaven Instruments Corp., Holtville, NY, USA) was found to be close to 33 nm. The resulted concentration of ferrofluid (5.1%) was determined as the dry weight residue after evaporation at 90 °C in an oven.

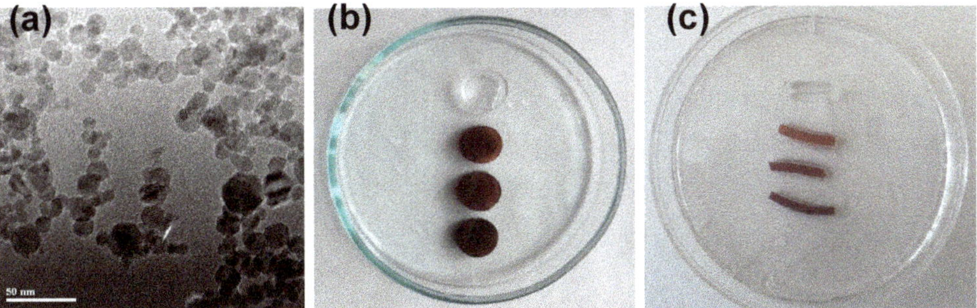

Figure 1. Transmission electron microscopy (TEM) image of laser target evaporation (LTE) iron oxide magnetic nanoparticles (MNPs) used for ferrogel synthesis (**a**). General view of gel and ferrogel samples from batch #1 (**b**) and batch #2 (**c**). See explanation in the text.

The stock ferrofluid was then diluted and used as a medium for the synthesis of FGs with varying content of MNPs. Monomer—acrylamide (AppliChem)—was used in 1.6 M concentration, a cross-linking agent—dimethylacrylamide (Merck)—was taken in 1/100 molar ratio to the monomer. Ammonium persulfate (APS) was used as an initiator in 3 mM concentration and the polymerization was performed at 70 °C for 2 h. N,N,N′,N′-tetramethylethylene diamine (TEMED) (Sigma Aldrich Inc. St. Louis MO, USA) in 6 mM concentration was used as a catalyst. After the synthesis, FGs were extensively washed in distilled water with daily water renewal for two weeks. During this period, FGs swelled to equilibrium. The equilibrium water uptake (Q) was determined by gravimetry while using the equation:

$$Q = \frac{m - m_0}{m_0},$$ (1)

with m denoting the weight of a swollen gel and m_0 denoting the weight of a residue after drying a gel in an oven at 70 °C. The values of Q were used for the calculation of the actual content of iron oxide MNPs in a swollen ferrogel (ω), according to the equation:

$$\omega = \frac{\gamma}{Q + 1},$$ (2)

with γ denoting the weight fraction of iron oxide MNPs in the dry residue of ferrogel. The value of γ was determined based on the composition of the reaction mixture in the synthesis.

Two batches of FG samples were synthesized differently in their shape. Ferrogels of batch #1 were synthesized in cylindrical polyethylene moulds 8.5 mm in diameter and 50 mm in height. These gels were used for ultrasonography studies in water, and further on used for the mechanical testing experiments. Therefore, they were cut into cylindrical plaques of the size of approximately 5 mm (in height). The diameter of cylinders was approximately 13 mm conditioned by the equilibrium swelling of a gel in water after the synthesis. FGs of batch #1 contained 0.00, 0.33, 0.64, and 1.34% of maghemite iron oxide MNPs by weight. Further on, they are denoted as FG1-0, FG1-1, FG1-2, and FG1-3 samples. A general view of FG samples of batch #1 is given in Figure 1b.

Ferrogels of batch #2 were synthesized in capillary polyethylene moulds 1.7 mm in diameter and 20 mm in height. These gels were used for ultrasonography studies in the configuration modelling the blood vessel (tube). These samples were cut in small cylinders of approximately 6 mm in length. The diameter of the cylinders was approximately 2 mm, provided by the equilibrium swelling of a gel in water after the synthesis. Ferrogels of batch #2 contained 0.00, 0.55, 0.98, and 1.45% of iron oxide MNPs by weight. Further on, they are denoted as FG2-0, FG2-1, FG2-2, and FG2-3 samples. Figure 2 provides a general view of ferrogel samples of batch #2. The content of iron oxide MNPs in ferrogels of batch#1 and batch#2 slightly differed due to the variation in the composition of the reaction mixtures. In the context of proposed study, we did not set out to synthesize samples with exact the same concentration of particles in two batches.

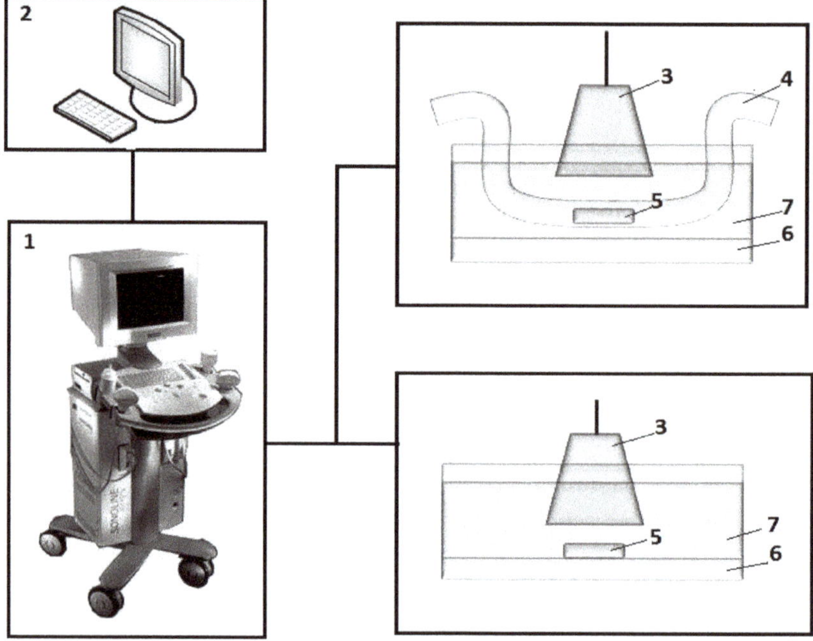

Figure 2. Scheme of the experimental setup to determine the echogenic properties of gels/ferrogels: 1—ultrasonic apparatus; 2—personal computer; 3—sensor of ultrasonic apparatus; 4—silicone tube with water; 5—sample of the gel/ferrogel; 6—soft pad; and, 7—cuvette with water. See also explanation in the text.

Synthesized ferrogels were uniformly colored and transparent (see Figure 1b,c). Visually, their appearance was similar to that of the precursor ferrofluid that was taken for the synthesis. No signs of turbidity were noticeable (including observations at the level of optical microscopy). Thus, it might be considered that there was no substantial aggregation of iron oxide MNPs in the synthesis of ferrogels and the distribution of MNPs in ferrogels was uniform, as in the precursor ferrofluid.

The elastic properties of the gels were evaluated by mechanical testing of the samples as described elsewhere [6,8,10]. Briefly, stepped strains for compression of up to 20% of the initial length were set to samples of gels while using a linear motor with a step of 1–2%. As a result, the "stress- strain" (σ-ε) dependencies were obtained, from the linear part of which the Young's modulus was determined for each one of the samples. For these series of experiments, samples of gels from the batch #1 (cylinders with a diameter of 13 mm and a height of 5 mm) were used (Figure 1b).

The magnetic hysteresis loops M(H) of the MNPs of iron oxide, gels, and ferrogels were measured by vibrating sample magnetometer (VSM, Faraday magnetometer of laboratory design). The maximum value of the magnetization was obtained for the external field of 1.5 kOe. For simplicity, it was denominated as saturation magnetization (Ms). The coercivity (Hc) was also estimated from the shape of the M(H) hysteresis loop. Thermomagnetic zero field cool–field cool curve (ZFC-FC) was measured for air dry MNPs following standard procedure for magnetic nanoparticles [19]. The external magnetic field for the ZFC-FC curve was as high as H = 100 Oe (it was applied for cooling and heating in the case of FC part of the curve and for heating only in the case of ZFC part of the curve).

2.2. Experimental Setup for the Gels' Ultrasound Visualization

Figure 2 shows an experimental setup for the study of the echogenic properties of gels and ferrogels. Samples of gels from the first batch (Figure 1b) were placed at the bottom of a cuvette with distilled water.

Samples from the second batch (Figure 1c) were located inside the silicone tube with an inner diameter of 6 mm and a wall thickness of 3 mm. The tube served for modelling the blood vessel configuration. It was filled with distilled water.

Fluid flow was not considered at this moment. Samples of the gels were visualized while using a Sonoline Adara (Siemens, Munich, Germany) medical device with a SIEMENS 7.5L45s Prima/Adara linear sensor. Ultrasonic sensor was immersed into a 500 mL cuvette that was filled with water for providing an acoustic contact. A gasket of soft viscose absorbent fibers was placed at the bottom of the cuvette in order to avoid the contribution of reflecting ultrasonic signal from the bottom of the cuvette. The video output of the ultrasound unit was connected to a computer equipped with an AverTV Hybrid VolarHX video capture device.

The dynamic range of the ultrasonic device in the mode of reception of the reflected oscillations was 66 dB, working frequency of 10 MHz, and the wavelength of 0.15 mm. In the experiments that were carried out within the framework of the corresponding batch of samples, the settings of the ultrasonic apparatus (radiation power, amplification, dynamic range, depth of visualization, etc.) were kept constant. The image of samples in two-dimensional (2D) mode in gray scale was recorded in a video file with a duration of several seconds with a frame rate of 25 frames per second, and the frame size is 720 pixels × 576 pixels.

Special software was developed to quantify the brightness of the image in various areas. It allowed to measure the brightness in the vicinity of a point specified by the user. In the experiment, the brightness of the image at the gel-water interface was measured and estimated. As a rule, the border thickness important for imaging was as high as 4–5 pixels. On this basis, the size of the area for assessing the brightness was limited to a square of 3 pixels × 3 pixels, which corresponds to linear dimensions of about 0.2 mm × 0.2 mm. For each sample, the measurements were carried out along the entire boundary of the gel. The number of measurements was at least 15 along the entire length of the sample. In each studied area, the minimum, maximum, and average image brightness were evaluated. The brightness was characterized in arbitrary units and it ranged from 0 (black) to 255 (white). For each type of FG, the average value of the maximum and average brightness, as well as the limits of the confidence interval at a significance level of $p = 0.05$, were calculated. In addition, in all experiments, 20 pixels × 20 pixels square image was used to estimate the background, i.e., the brightness of the image area where the water was located. The maximum, minimum and average background brightness in all cases varied insignificantly. Therefore, finally, the adjustment that was associated with changes in the brightness of the background was not carried out.

3. Results and Discussion

3.1. Structural and Magnetic Characterization of Nanoparticles and Ferrogels

Figure 1a shows an example of transmission electron microscopy (TEM) image of iron oxide MNPs (JEOL JEM2100, Tokyo, Japan). The majority of MNPs have spherical shape. Very few of them contain hexagonal corners. Particle size distribution (PSD) of MNPs was lognormal with the median $d_0 = 11.7$ nm and the logarithmic dispersion $\sigma = 0.423$, as determined by the graphical analysis of 2150 TEM images. According to PSD, the average diameter of 93% of MNPs (by weight) fits a 5–40 nm range. According to X-ray diffraction (Bruker D8 Discover, Billerica, MA, USA), the crystalline structure of MNPs was an inverse spinel with space group Fd3m. The lattice parameters corresponded to maghemite (Fe_2O_3). The oxidation number +3 of Fe ions in the chemical composition of MNPs was confirmed by Ox/Red titration (TitroLine, Schott Instruments).

Measurements of the saturation magnetization of as-prepared air dried MNPs showed that their average size and defined composition (γ-Fe$_2$O$_3$) are quite consistent with each other: $M_s \approx 40$ emu/g. Detailed discussion on the magnetic structure of "core-shell" LTE MNPs can be found elsewhere [19,20].

As before, pure gel without nanoparticles showed linear non-hysteretic diamagnetic response on the application of external magnetic field (Figure 3a). At the same time, the ferrogel's M(H) loops had a typical S-shape with negligible coercivity for the small concentrations of MNPs (Figure 3b). The evolution of the Ms value as a function of the concentration of nanoparticles in the ferrogel shows linear dependence: the higher concentration, the higher the Ms (Figure 3b). The magnetization value for the FG scaffold in the applied magnetic field of certain strength is a very important parameter in a view of FGs applications for drug delivery and the controlled movement of micro-objects by magnetic field. For the highest concentration of 1.45 wt. % in the external field of 0.5 kOe (reasonably low an accessible for generation), the magnetic moment of the FG was as high as 0.5 emu/g. Taking into account the scaffold FG2-3 volume, one can obtain the magnetic moment of the sample in the 0.5 kOe external magnetic field: m = 0.9 × 10^{-3} emu.

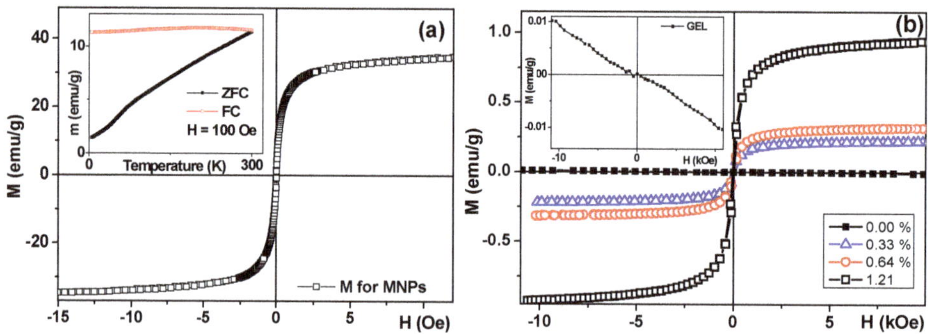

Figure 3. Magnetic hysteresis loop of air-dry nanoparticles measured at room temperature; inset shows thermomagnetic zero field cool–field cool curve (ZFC-FC) curve (**a**). Hysteresis loops for FG1-0, FG1-1, FG1-2, and FG1-3 samples measured at room temperature (**b**); inset shows diamagnetic response of the blank gel.

3.2. Gels Elasticity

Figure 4a shows a plot of 'stress-strain' relationship for one of the samples of the first batch of gels (0.00, 0.33, 0.64 and 1.34 wt.% of MNPs). It is seen that, at any fixed strain, the stress in the gel is the greater, the higher the concentration of MNPs. The caption contains the linear regression equations for each sample, where the value of ε corresponds to the value of Young's modulus in kPa.

Figure 4b shows the dependence of the Young's modulus on the concentration of MNPs in the gel/ferrogel for all the tested samples (six samples for each type of gel/ferrogel). First of all, it can be seen that the addition of MNPs to the PAA gel in the minimum concentration (0.33%) leads to a significant increase in the Young's modulus of the composite material. Secondly, a gradual increase in the concentration of MNPs in ferrogel is accompanied by a further increase in its elasticity. The results that were obtained are in good agreement with the data of our earlier studies [6,8,11] and the findings of other authors [7,21].

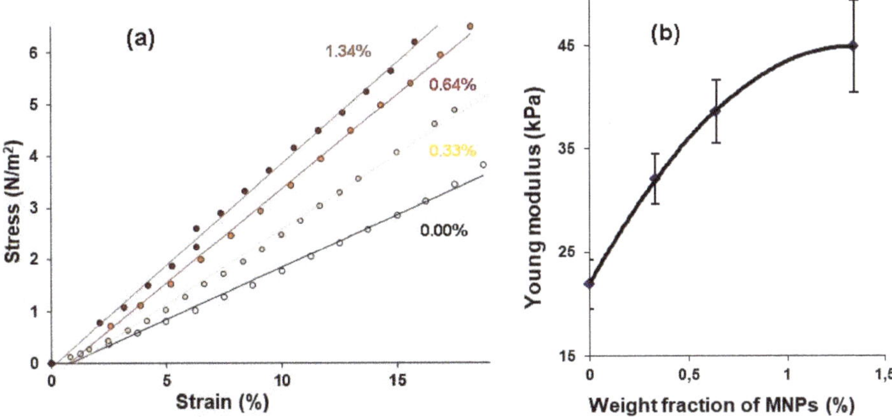

Figure 4. Example of the 'stress-strain' relationship for gels of batch #1 (**a**) and the dependence of Young's modulus on the MNPs concentration in gels of batch #1 (**b**). Vertical bars reflect the confidence interval with p = 0.05. Following fit parameters were obtained for stress (σ). For concentration 0.00%: $\sigma = 20.2\varepsilon - 0.2$, $R^2 = 0.995$. For concentration 0.33%: $\sigma = 28.9\varepsilon - 0.3$, $R^2 = 0.994$. For concentration 0.64: $\sigma = 36.4\varepsilon - 0.3$, $R^2 = 0.997$. For concentration 1.34: $\sigma = 39.2\varepsilon - 0.1$, $R^2 = 0.998$.

3.3. Gel/water Boundary Echogenicity

Figure 5 shows an example of a single ultrasound image frame when scanning a gel sample positioned at the bottom of a cuvette (a) and inside a model vessel (b) (Figure 2). The images very clearly show the boundaries of the gel samples lining at the bottom of the cuvette and the walls of the tube with water. It is also seen that the highest echogenicity of the gels (i.e., the largest amplitude of the reflected ultrasonic vibrations, and, accordingly, the brightness of the image) corresponds not to the sample body, but to the interface between the gel surface and the water. In experiments with the first batch of gels, the visualization was performed with an ultrasound device amplification of 15 dB. Four samples were tested for each concentration of MNPs in the gel. The average value of the maximum brightness of the water in the cuvette in all tests of the gels was 34.1 ± 0.2 (n = 16).

Figure 6a shows the dependence of the maximum and the average brightness at the gel/water interface on the concentration of MNPs in the gel. The graphs show the boundaries of confidence intervals at p = 0.05. It can be seen that, by increasing the weight fraction of the MNPs in the samples, both the maximum brightness (1) and the average brightness (2) of the echo reflected from the surface of the gels were increased. The data are well approximated by the linear regressions: (1) y = 16.783x + 205.4, $R^2 = 0.988$; (2) y = 8.166x + 183.8, $R^2 = 0.931$.

Figure 6b shows the dependences of the maximum and average brightness of the gel/water interface on the Young's modulus of the samples. It can be seen that the echogenicity of the surface of the gels is directly related to the elastic properties of the tested materials. Moreover, the higher the stiffness of the samples, the more accurately the gels are visualized. Again, the data (Figure 6b) are well approximated by the linear regressions: (1) y = 1.529x + 169.45, $R^2 = 0.989$; and, (2) y = 0.7666x + 165.63, $R^2 = 0.987$.

In a series of experiments with the second batch of gels (samples were placed in a model "blood" vessel (Figures 1c and 5b), the ultrasonic amplification of the apparatus was increased to 20 dB to compensate for the signal loss through the tube wall. Totally, seven samples were tested for each concentration of MNPs in the gel. The average brightness of the water inside the tube was 44.4 ± 0.2 (n = 28).

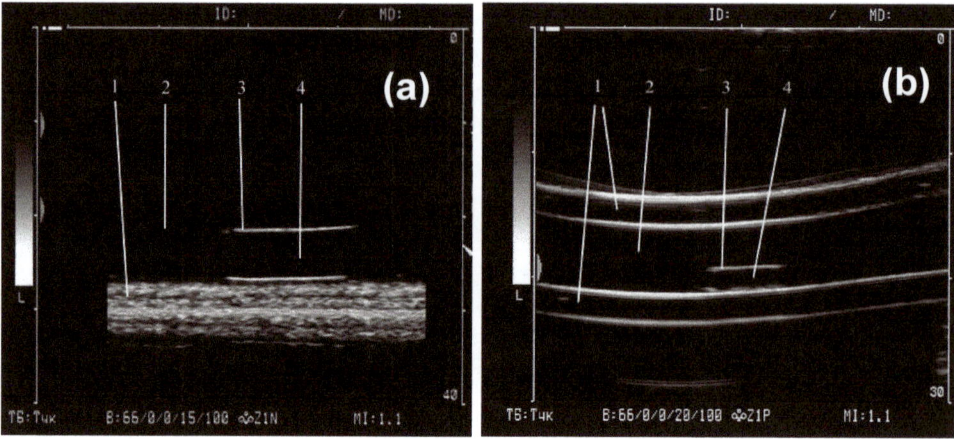

Figure 5. Examples of ferrogels samples visualization at the bottom of water cuvette (**a**) and inside a silicone tube filled with water (**b**). The distance from sensor to objects is about 20 mm. 1—soft pad (a) and wall of silicone tube (b); 2—water in cuvette (a) and in tube (b); 3—upper boundary of gel/water; 4—gel body.

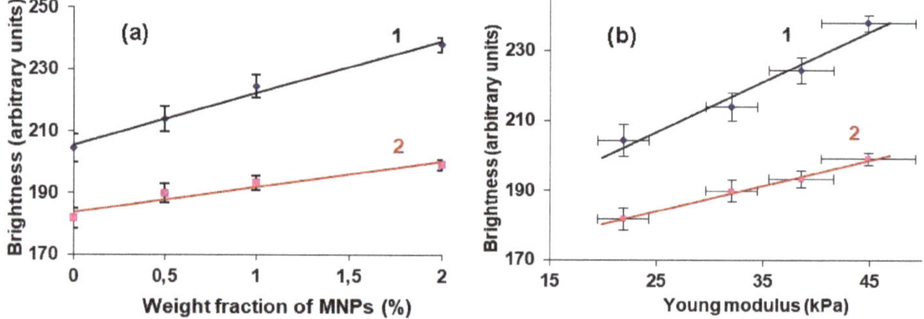

Figure 6. Dependences of maximal (1) and average brightness (2) at gel/water boundary on the MNPs concentration for samples of batch #1 (**a**). Vertical bars reflect the confidence interval with p = 0.05. Dependences of maximum (1) and average brightness (2) at gel/water boundary on Young's modulus for samples of batch #1 (**b**). Vertical and horizontal bars reflect the confidence interval of relevant parameter with p = 0.05.

Figure 7 shows a graph of the maximum and average brightness of the reflected echo signal at the gel/water interface in a tube versus the concentration of MNP in the sample. The graphs show the boundaries of confidence intervals at p = 0.05. The data (Figure 7) are well approximated by the linear regressions: (1) y = 61.652x + 130.07, R^2 = 0.996; (2) y = 31.9x + 81.6, R^2 = 0.984. It can be seen that the echogenicity of the surface of the ferrogel significantly increases in comparison with the PAA gel, even with the minimum concentration of MNPs. At the maximum concentration of MNPs in ferrogel (1.45%), the brightness of the reflected echo signal from its boundary with water is approximately two times higher than for the baseline PAA of the sample. This conclusion is valid for both the use of maximum and average brightness as a measure of the echogenicity of the material. The result that was obtained for the samples of gels (the first batch) inside the model vessel is in full compliance with the test data of the echogenicity of the samples (second batch) in water (see Figure 6a).

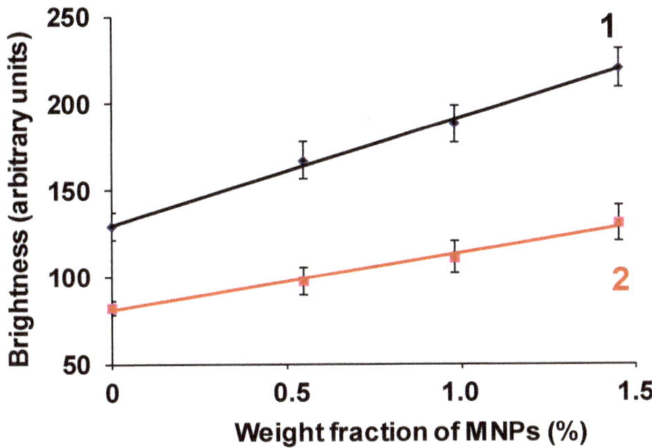

Figure 7. Dependences of maximum (1) and average brightness (2) at gel/water boundary on the MNPs concentration for samples of batch #2. Vertical bars reflect the confidence interval with $p = 0.05$.

It should be noted that the dependence of brightness at the gel/water interface on the concentration of MNPs is more pronounced for the samples in a model vessel than for gels that were placed in a cuvette filled with water.

We have studied the ultrasound reflection from the non-deformed sample in condition, which mimic the ferrogel in a blood vessel. It is important to mention that effect of the sample preliminary tension on the sound reflection is out of scope of this work. Figures 4, 6 and 7 demonstrate that the Young modulus and the brightness of the reflected signal both increase with the particle concentration. It allows for making simple comment on the relation between the Young's modulus and the impedance and to conclude that the impedance increases with the modulus. Note that classical results of the acoustic theory demonstrate the increasing relation on the basis of the general considerations of the theory of continuous media.

4. Discussion

The experience of ultrasound imaging of such polymer soft materials as gels for medical purposes is well known. In particular, the results of the location of gel phantoms of internal organs and tissues were used for testing and calibration of ultrasound diagnostic equipment [22,23], as well as implants that are based on gels [24,25]. Here, the visualization of FG sample geometrical features was for the first time achieved by using the ultrasonic location method. While taking into account the fact that FGs are considered to be promising materials for medical applications, the study was performed by a standard ultrasound apparatus that is in use for medical purposes. Regardless the shape of the samples and the conditions of their location, the boundaries of ferrogels are confidently visualized while using medical ultrasound over the entire range of concentrations of MNPs.

An increase in the MNPs concentration in the PAA gel was accompanied not only by an increase in the echogenicity of the gel/water interface, but also by the elasticity of the samples. Moreover, we established a linear relationship between the brightness of the echo signal reflected from the boundary and the Young's modulus of the FG (Figure 6b). This result was obtained for the samples from the first batch of gels, while visualizing them at the bottom of a container with water.

It should be mentioned that in a series of experiments with a model vessel, the elastic properties of FGs from the second batch were not determined due to the design of the equipment for mechanical properties evaluation. When the diameter of the samples was below 8 mm, the installation did not allow for correct setting of deformations in compression. However, all of the materials and procedures used for the synthesis of gels were the same. The only exception was the moulding of gels for relevant needs. This circumstance makes it possible to insure the possible connection between the echogenicity of the surface of the gels and the Young's modulus of samples from the second series with a high degree of confidence. Below, we attempt to describe the relationship between the concentration of MNPs in the gel and its echogenicity at the interface with water theoretically.

In this part, we suggest some physical interpretation of the experimental results. It is based on the concept that, at the acoustic oscillations of the hydrogel, the macromolecular net and water in the gel porous experience the viscous Stocks-like interaction. Thus, in the first approximation, one can suppose that, in each small volume of the hydrogel, the net and water move with the same velocity and the hydrogel can be considered as a unite continuum. In the frame of this physical approximation, the coefficient R of the sound reflection on the border between hydrogel and water out of it, can be estimated, as follows [26]

$$R = \left(\frac{c_2\rho_2 - c_1\rho_1}{c_2\rho_2 + c_1\rho_1} \right)^2 \tag{3}$$

Here, the Subscribes 1 and 2 correspond to the water out of the hydrogel and the hydrogel, considered as a continuous medium, respectively, ρ is mass density of the corresponding media.

The sound speed can be presented as in Ref. [26]

$$c = \sqrt{\frac{K}{\rho}} \tag{4}$$

where K is the compression modulus of the media. Therefore, Equation (3) can be rewritten as form:

$$R = \left(\frac{\sqrt{K_2\rho_2} - \sqrt{K_1\rho_1}}{\sqrt{K_2\rho_2} + \sqrt{K_1\rho_1}} \right)^2 \tag{5}$$

The hydrogel compression modulus K_2 is determined by the modulus of the water, modulus of the macromolecular net, and modulus of the embedded particles (see, for example, general discussion in [27]. Since the volume concentration of the gel and particles in the hydrogel are low, in the first approximation $K_2 \approx K_1$. The hydrogel mass density is more than the density of water, i.e., $\rho_2 > \rho_1$. Therefore, the inequality $(K_2\rho_2)^{1/2} > (K_1\rho_1)^{1/2}$ is held and the coefficient R can have quite significant value, which is enough to provide the visible reflection signal (see Figure 4). The addition of the rigid particles increases both the modulus K_2 of the hydrogel and its density ρ_2. Obviously, in the range of small concentrations of the particles, coefficient R (defined for Equation (5)) should show linear dependence on the concentration (Figures 6a and 7).

The suggested theoretical interpretation of the experimental results is rather qualitative. Since the ultrasound wavelength is much more than the size of the particles and heterogeneities, provoked by the particles, we have used classic considerations of the acoustics of continuous media. It appeared enough for, at least, principal explanation of the experimental results. Detailed experimental study of the particles distribution inside the hydrogel are required in order to achieve reliable quantitative description of the reflection effects at higher level of understanding.

Let us make short remark on possible future directions of ultrasound location research. As it was mentioned in the introduction, magnetically controlled platforms for targeted delivery of cell implants and drug substrates through arteries, as well as positioning indicators for low invasive surgery, is a hot topic of the research and applications. Recently, we proposed using giant magnetoimpedance based multilayered sensitive element for the monitoring of FG scaffold position inside the blood vessel in regenerative medicine case of application [8,28].

Generally speaking, the goal to define the FG scaffold position can be completed either by magnetic or by ultrasound detection. What is most important is that both techniques can probably be used at a time or just one after another in two simple non-invasive tests. The concept of usage of different techniques for material characterization is well established in the nanomedicine for nanomaterials characterization [16]. Ferrogels offer a new opportunity for extending this concept on the non-invasive tests, keeping in mind not a material characterization, but rather complex diagnostic solution. Different techniques have their advantages and disadvantages and the application of multiple techniques can be viewed as additional advantage, especially keeping in mind than ultrasonic equipments are available for routine diagnostics.

5. Conclusions and Outlook

Samples of ferrogels with different concentrations of MNPs were synthesized in various forms to simulate different peculiarities of the visualization of sample location for future development in the area of biomedical applications. In particular, thin cylindrical samples were used to simulate the situation in which ferrogels can be used as platforms for targeted drug delivery or cell cultures through blood vessels. Samples of larger size in the form of cylindrical plates were used in order to simulate cell grafts that are based on ferrogel, intended for use as substitutes for damaged structures, such as bone or cartilage tissues.

We found that, regardless of the shape of the samples and the conditions of their location while using medical ultrasound, the boundaries of ferrogels are confidently visualized over the entire range of concentrations of MNPs. Moreover, the intensity of the reflected echo was greater, the greater the concentration of MNPs in the gel. Obviously, the result that was obtained is not related to the influence of the MNPs directly on the intensity of the reflected echo signal, since the wavelength of the ultrasonic effect used is much larger than the particle size. Therefore, we can point out the indirect effect of MNPs on the echogenicity of ferrogels.

Qualitative theoretical interpretation of the experimental results was proposed while taking into account the concept that, at the acoustic oscillations of the hydrogel, the macromolecular net and water in the gel porous experience the viscous Stocks-like interaction.

Author Contributions: F.A.B., S.Y.S., A.P.S. and G.V.K. conceived and designed the experiments; F.A.B., S.Y.S., O.A.D., T.F.S. and A.P.S. performed the experiments; S.Y.S. developed the special software; F.A.B., O.A.D., T.F.S., A.P.S. and G.V.K. analyzed the data; A.Y.Z. proposed the theoretical justification of data obtained; F.A.B., A.P.S., A.Y.Z. and G.V.K. wrote the manuscript. All authors discussed the results and implications, and commented on the manuscript at all stages. All authors read and approved the final version of the manuscript.

Funding: The Russian Scientific Foundation (grant 18-19-00090) supported the experimental parts of this study, including the design, performance and analysis of experiments.

Acknowledgments: A.Yu. Zubarev thanks the program of the Ministry of Education and Science of the Russian Federation (project 3.1438.2017/46) for the support of his mathematical studies. We thank K.R. Mekhdieva, P.A. Shabadrov, V.Ya. Krokhalev, I.V. Beketov and A.M. Murzakaev for special support.

Conflicts of Interest: The authors declare no conflict of interest.

References

1. DeRossi, D.; Kajiwara, K.; Osada, Y.; Yamauchi, A. *Polymer Gels: Fundamentals and Biomedical Applications*; Plenum Press: New York, NY, USA, 1991.
2. Filipcsei, G.; Csetneki, I.; Szilágyi, A.; Zrínyi, M. Magnetic Field-Responsive Smart Polymer Composites. *Adv. Polym. Sci.* **2007**, *206*, 137–189. [CrossRef]
3. Li, Y.; Huang, G.; Zhang, X.; Li, B.; Chen, Y.; Lu, T.; Lu, T.; Xu, F. Magnetic Hydrogels and their potential biomedical applications. *Adv. Func. Mater.* **2013**, *23*, 660–672. [CrossRef]
4. Safronov, A.P.; Mikhnevich, E.A.; Lotfollahi, Z.; Blyakhman, F.A.; Sklyar, T.F.; Varga, A.L.; Medvedev, A.I.; Armas, S.F.; Kurlyandskaya, G.V. Polyacrylamide Ferrogels with Magnetite or Strontium Hexaferrite: Next Step in the Development of Soft Biomimetic Matter for Biosensor Applications. *Sensors* **2018**, *18*, 257. [CrossRef] [PubMed]
5. Kennedy, S.; Roco, C.; Délérisa, A.; Spoerri, P.; Cezar, C.; Weaver, J.; Vandenburgh, H.; Mooney, D. Improved magnetic regulation of delivery profiles from ferrogels. *Biomaterials* **2018**, *161*, 179–189. [CrossRef] [PubMed]

6. Blyakhman, F.A.; Safronov, A.P.; Zubarev, A.Y.; Shklyar, T.F.; Makeyev, O.G.; Makarova, E.B.; Melekhin, V.V.; Larrañaga, A.; Kurlyandskaya, G.V. Polyacrylamide ferrogels with embedded maghemite nanoparticles for biomedical engineering. *Results Phys.* **2017**, *7*, 3624–3633. [CrossRef]

7. Bonhome-Espinosa, A.B.; Campos, F.; Rodriguez, I.A.; Carriel, V.; Marins, J.A.; Zubarev, A.; Duran, J.D.G.; Lopez-Lopez, M.T. Effect of particle concentration on the microstructural and macromechanical properties of biocompatible magnetic hydrogels. *Soft Matter* **2017**, *13*, 2928–2941. [CrossRef] [PubMed]

8. Blyakhman, F.; Buznikov, N.; Sklyar, T.; Safronov, A.; Golubeva, E.; Svalov, A.; Sokolov, S.; Melnikov, G.; Orue, I.; Kurlyandskaya, G. Mechanical, electrical and magnetic properties of ferrogels with embedded iron oxide nanoparticles obtained by laser target evaporation: Focus on multifunctional biosensor applications. *Sensors* **2018**, *18*, 872. [CrossRef]

9. Lopez-Lopez, M.T.; Scionti, G.; Oliveira, A.C.; Duran, J.D.G.; Campos, A.; Alaminos, M.; Rodriguez, I.A. Generation and Characterization of Novel Magnetic Field-Responsive Biomaterials. *PLoS ONE* **2015**, *10*, e0133878. [CrossRef]

10. Blyakhman, F.A.; Safronov, A.P.; Makeyev, O.G.; Melekhin, V.V.; Shklyar, T.F.; Zubarev, A.Y.; Makarova, E.B.; Sichkar, D.A.; Rusinova, M.A.; Sokolov, S.Y.; et al. Effect of the polyacrylamide ferrogel elasticity on the cell adhesiveness to magnetic composite. *J. Mech. Med. Boil.* **2018**, *18*, 1850060. [CrossRef]

11. Blyakhman, F.A.; Makarova, E.B.; Fadeyev, F.A.; Lugovets, D.V.; Safronov, A.P.; Shabadrov, P.A.; Shklyar, T.F.; Melnikov, G.Y.; Orue, I.; Kurlyandskaya, G.V. The Contribution of Magnetic Nanoparticles to Ferrogel Biophysical Properties. *Nanomaterials* **2019**, *9*, 232. [CrossRef]

12. Józefczak, A.; Kaczmarek, K.; Kubovčíková, M.; Rozynek, Z.; Hornowski, T. The effect of magnetic nanoparticles on the acoustic properties of tissue mimicking agar-gel phantoms. *J. Magn. Magn. Mater.* **2016**, *431*, 172–175. [CrossRef]

13. Kaczmarek, K.; Hornowski, T.; Dobosz, B.; Józefczak, A. Influence of magnetic nanoparticles on the focused ultrasound hyperthermia. *Materials* **2018**, *11*, 1607. [CrossRef] [PubMed]

14. Zverev, V.I.; Pyatakov, A.P.; Shtil, A.A.; Tishin, A.M. Novel applications of magnetic materials and technologies for medicine. *J. Magn. Magn. Mater.* **2018**, *459*, 182–186. [CrossRef]

15. Zamora-Mora, V.; Soares, P.I.P.; Echeverria, C.; Hernández, R.; Mijangos, C. Composite chitosan/agarose ferrogels for potential applications in magnetic hyperthermia. *Gels* **2015**, *1*, 69–80. [CrossRef] [PubMed]

16. Kim, C.; Kim, H.; Park, H.; Lee, K.Y. Controlling the porous structure of alginate ferrogel for anticancer drug delivery under magnetic stimulation. *Carbohydr. Polym.* **2019**, *223*, 115045. [CrossRef] [PubMed]

17. Liu, T.; Hu, S.; Liu, T.; Liu, D.; Chen, S. Magnetic-sensitive behaviour of intelligent ferrogels for controlled release of drug. *Langmuir* **2006**, *22*, 5974–5978. [CrossRef] [PubMed]

18. Grossman, J.H.; McNeil, S.E. Nanotechnology in cancer medicine. *Phys. Today* **2012**, *65*, 38–42. [CrossRef]

19. Safronov, A.P.; Beketov, I.V.; Komogortsev, S.V.; Kurlyandskaya, G.V.; Medvedev, A.I.; Leiman, D.V.; Larranaga, A.; Bhagat, S.M. Spherical magnetic nanoparticles fabricated by laser target evaporation. *AIP Adv.* **2013**, *3*, 052135. [CrossRef]

20. Safronov, A.P.; Beketov, I.V.; Tyukova, I.S.; Medvedev, A.I.; Samatov, O.M.; Murzakaev, A.M. Magnetic nanoparticles for biophysical applications synthesized by high-power physical dispersion. *J. Magn. Magn. Mater.* **2015**, *383*, 281–287. [CrossRef]

21. Noorjahan; Pathak, S.; Jain, K.; Pant, R. Improved magneto-viscoelasticity of cross-linked PVA hydrogels using magnetic nanoparticles. *Colloids Surf. A Physicochem. Eng. Asp.* **2018**, *539*, 273–279. [CrossRef]

22. Zell, K.; Sperl, J.I.; Vogel, M.W.; Niessner, R.; Haisch, C. Acoustical properties of selected tissue phantom materials for ultrasound imaging. *Phys. Med. Biol.* **2007**, *52*, N475–N484. [CrossRef] [PubMed]

23. Lafon, C.; Zderic, V.; Noble, M.L.; Yuen, J.C.; Kaczkowski, P.J.; Sapozhnikov, O.A.; Chavrier, F.; Crum, L.A.; Vaezy, S. Gel phantom for use in high-intensity focused ultrasound dosimetry. *Ultrasound Med. Boil.* **2005**, *31*, 1383–1389. [CrossRef] [PubMed]

24. Othman, N.S.; Jaafar, M.S.; Rahman, A.A.; Othman, E.S.; Rozlan, A.A. Ultrasound Speed of Polymer Gel Mimicked Human Soft Tissue within Three Weeks. *Int. J. Biosci. Biochem. Bioinform.* **2011**, *1*, 223–225. [CrossRef]

25. Vaezy, S.; Noble, M.L.; Kaczkowski, P.J.; Martin, R.W.; Crum, L.A. Polyacrylamide gel as an acoustic coupling medium for focused ultrasound therapy. *Ultrasound Med. Biol.* **2003**, *29*, 1351–1358.

26. Landau, L.; Lifshitz, E. *Fluid Mechanics*; Pergamon Press: Oxford, UK, 1987.

27. Christensen, R.M. *Mechanics of Composite Materials*; Krieger Publishing Company: Malabar, FL, USA, 1991.
28. Kurlyandskaya, G.V.; Fernández, E.; Svalov, A.; BurgoaBeitia, A.; García-Arribas, A.; Larrañaga, A. Flexible thin film magnetoimpedance sensors. *J. Magn. Magn. Mater.* **2016**, *415*, 91–96. [CrossRef]

 *sensors*

Article

Size-Dependent Properties of Magnetosensitive Polymersomes: Computer Modelling

Aleksandr Ryzhkov [1,*,†] **and Yuriy Raikher** [2,†]

[1]   Laboratory of Mechanics of Functional Materials, Institute of Continuous Media Mechanics, Ural Branch, Russian Academy of Sciences, 614068 Perm, Russia
[2]   Laboratory of Physics and Mechanics of Soft Matter, Institute of Continuous Media Mechanics, Ural Branch, Russian Academy of Sciences, 614068 Perm, Russia; raikher@icmm.ru
*   Correspondence: ryzhkov.a@icmm.ru
†   These authors contributed equally to this work.

Received: 22 October 2019; Accepted: 26 November 2019; Published: 29 November 2019

Abstract: Magnetosensitive polymersomes, which are amphiphilic polymer capsules whose membranes are filled with magnetic nanoparticles, are prospective objects for drug delivery and manipulations with single cells. A molecular dynamics simulation model that is able to render a detailed account on the structure and shape response of a polymersome to an external magnetic field is used to study a dimensional effect: the dependence of the field-induced deformation on the size of this nanoscale object. It is shown that in the material parameter range that resembles realistic conditions, the strain response of smaller polymersomes, against *a priori* expectations, exceeds that of larger ones. A qualitative explanation for this behavior is proposed.

Keywords: magnetic polymersomes; magnetic vesicles; magnetic nanoparticles; magnetoactive composites; nanocapsules; coarse-grained molecular dynamics; computer simulation

1. Introduction

Synthesis of microscale particles capable of drug transportation and/or active manipulations with cells is one of the most actively developing trends in modern biomedicine [1]. A promising class of such objects, well tunable and controllable, are polymersomes: vesicle-like submicron capsules whose walls are formed by a bilayer membrane built of an amphiphilic block copolymer [2–7]. The polymersome might be loaded with biomedical substances (cargo) whereas the membrane might be functionalized to be stimuli-sensitive to a number of factors: pH, temperature, oxidation/reduction, electric/magnetic fields, light, ultrasound, etc. Chemical versatility and stability of polymersomes ensure a variety of their foreseen applications: biosensing systems, nanoreactors, drug carriers with externally activated release, imaging, and even prototypes of artificial cell organelles.

The objects of current study are magnetosensitive polymersomes (magnetopolymersomes, MPSs) whose membrane is modified by introducing inside it magnetic (e.g., maghemite) nanoparticles (MNPs) [8–11]. Due to the hydrophilic/hydrophobic interplay, the MNPs are confined inside the sandwich structure formed by the inner and outer amphiphilic polymer layers (shells); the width of the intershell gap is adjusted in such a way that it just a bit exceeds the MNP size. With such a construction, magnetodeformational susceptibility of MPSs comes out several orders of magnitude higher than that of any other types of microcapsules.

Nowadays, chemical synthesis of various modifications of MPSs is rapidly developing, and they are available in different shapes and morphologies [12,13]. In [14], membranes of poly(trimethylene carbonate)-*b*-poly(L-glutamic acid) (PTMC-*b*-PGA) with incorporated maghemite ($\gamma - Fe_2O_3$) nanoparticles are reported. The monolayer, double-layer and multilayer vesicles with poly(styrene)-*b*-poly(acrylic acid) (PS-*b*-PAA) membranes of tuned thickness and stuffed with MNPs are obtained in [15]. The tests demonstrate that the polymersomes with thicker membrane and higher MNP density display enhanced MRI contrast, higher magnetization and better release profile of their drug cargo. In structure investigations, it is found that larger MNPs drift closer to the inner membrane boundary that leads to larger curvature of the latter [16]. Besides dilute systems, preparation of size-controlled assemblies of densely packed submicron PS-*b*-PAA polymersomes is also possible [16].

Now a commonly established fact is that under an applied external field, MPSs notably stretch along the field direction. Moreover, as revealed by small-angle neutron scattering (SANS), the field causes changes in the membrane structure making it anisotropic, and this anisotropy increases with the applied field. The evaluated scattering anisotropy factor is also found to be dependent on the MNP content, size and curvature of the membrane. According to the SANS observations, the field strength at which the field-dependence of the anisotropy factor attains a plateau coincides with that saturating the MNP magnetic moments [14].

The granted biocompatibility of MPSs and experimentally proven opportunity to magnetically control their shape as well as the changes of MNP distribution inside the membrane, commend them as really smart micro-objects for biomedical use. For further progress, one needs a valid concept of complex magnetomechanical and magnetostructural properties of MPSs that, as for now, is virtually nonexistent. A promising, if not the only, way to fill in this gap, i.e., to study and predict the deformational and structure responses of MPSs, is an extensive use of computer simulations. In what follows, we describe a flexible computational model of an MPS and use it to analyze the field-induced shape and volume effects.

2. Model and Simulations

A coarse-grained molecular dynamics description of a magnetic polymersome was developed in [17]. Here, we use it to investigate the dependence of magnetodeformational behavior of an MPS on the overall size of the latter.

The model MPS is built of the particles of two types: non-magnetic polymer ones (beads) and MNPs. The sandwiched amphiphilic membrane, within which the MNPs are confined, is presented as two nested spherical shells, the inner and outer ones, see Figure 1. Each shell consists of an equal number of beads arranged in a quasi-2D triangular mesh, so that the centers of beads are the mesh nodes; the diameters of the beads are denoted as d_{in} and d_{out}, respectively, where the subscript indicates the shell.

Inside each shell the bead centers are connected with stiff springs to ensure virtual constancy of the inter-bead distances and, thus, the area of the triangle built of any three neighboring beads. At the surface organized in such a way, any pair of side-abutting triangles is, however, free to fold along their mutual border. In other words, the overall surface area of the shell is conserved but the shell bending elasticity is negligible.

Inside this double-shell membrane, in a number of points uniformly distributed over its surface, the subtending beads (nodes) of the shells are linked with identical linear elastic springs; the stress-free length h of the spring defines the equilibrium thickness of the intershell layer. Given that the total membrane thickness is $h + \frac{1}{2}(d_{in} + d_{out})$, whereas the gap accessible for the MNPs is $h_{gap} = h - \frac{1}{2}(d_{in} + d_{out})$. The connectivity parameter c_b of the membrane is defined as the ratio of the number of spring-bonded pairs to the whole number of particles in the shell; evidently, diminution of c_b enhances the extent of the layer thickness fluctuations.

Inside the MPS membrane, a certain number of monodisperse spherical magnetic nanoparticles is confined, thus making a kind of quasi-2D magnetic fluid. The volume fraction ϕ of the MNPs is defined with respect to the volume of the intershell space where they are allowed to move. From the calculation viewpoint, each MNP is treated as a structureless bead of diameter d_p with a built-in magnetic (dipole) moment fixed at its center. The intershell gap h_{gap} but slightly exceeds d_p, so that the intershell layer might accommodate only a monolayer of MNPs. In magnetic aspect, the MNPs are coupled with each other as point dipoles; in steric aspect, they interact as soft spheres. The diameters d_{in} and d_{out} of the shell beads are chosen is such a way that the shells are impenetrable for the MNPs.

To commence simulation, first, under the above-presented conditions a multi-element object (a model MPS) is constructed as a hollow sphere of outer diameter D with a double-shell wall, within which a given number of MNPs is uniformly distributed; the number of the beads in the shells is chosen accordingly. Then all the interparticle interactions are "switched on", and the model MPS is set in contact with Langevin thermostat that induces Brownian motion of all the elements of the system. Therefore, the thermostat imitates the presence of isothermal solvent (consisting of light-weight molecules) which uniformly fills in all the space outside and inside the MPS.

From this instant the coarse-grained molecular dynamics calculation starts and is carried out until the system comes to equilibrium; the criterion for the latter is stabilization of the overall energy value. This stage yields the basic (field-free) state of the MPS and enables one to obtain its characteristics. Then a uniform magnetic field is imposed, i.e., the Zeeman energy is added to the energy of each magnetic element, and the calculation process is run anew.

Technically, each calculation implies the numerical solution of the set of coupled equations of motion for all the center-of-mass position vectors $\vec{r}_i\,(t)$ where subscript i enumerates all the elements of the model MPS, beads or magnetic particles:

$$m_i \frac{d^2\vec{r}_i}{dt^2} = \vec{F}_i - \zeta\frac{d\vec{r}_i}{dt} + \vec{f}_i\,(t)\,; \tag{1}$$

here m_i is the element mass, $\vec{F}_i$ the force derived from the total potential energy. The second term in the right-hand side of Equation (1) is the dissipative force with translational friction coefficient ζ, and $\vec{f}_i\,(t)$ is the random force generated by Langevin thermostat.

The force experienced by each bead of any shell is

$$\vec{F}_i = \vec{F}_{s,i} + \vec{F}_{a,i} + \vec{F}_{\text{bond},i} + \vec{F}_{\text{MNP-membrane},i},$$

with $\vec{F}_{s,i}$ being the total stretching/compression force on the part of neighboring mesh nodes, $\vec{F}_{a,i}$ a sum of forces depending on relative change of the areas of the mesh triangles which have the i-th node as a common vertex, $\vec{F}_{\text{bond},i}$ an elastic force entailed by the presence of the bonds connecting the i-th node of one shell with the subtending node of another one, $\vec{F}_{\text{MNP-membrane},i}$ a total force of soft mutual repulsion between the i-th and all the other MNPs, this set of forces is defined by pairwise Weeks-Chandler-Andersen (truncated Lennard-Jones) potential [18].

The force acting on each MNP is

$$\vec{F}_i = \vec{F}_{\text{dipolar},i} + \vec{F}_{\text{MNP-MNP},i} + \vec{F}_{\text{MNP-membrane},i}. \tag{2}$$

Here $\vec{F}_{\text{dipolar},i}$ is the sum of pairwise forces derived from the dipolar magnetic interaction of the i-th MNP that bears magnetic moment μ directed along unit vector $\vec{e}_i$ with all the others located at the intercenter distances $\vec{r}_{ij}$. This force has the form

$$-\nabla\left[\mu_0\mu^2\left(\frac{(\vec{e}_i\cdot\vec{e}_j)}{r_{ij}^3} - \frac{3\,(\vec{e}_i\cdot\vec{r}_{ij})\,(\vec{e}_j\cdot\vec{r}_{ij})}{r_{ij}^5}\right)\right]; \tag{3}$$

here μ_0 is vacuum permeability. The second and third terms in the right-hand part of Equation (2) are the soft repulsion forces between MNPs and between MNPs and the confining shells, respectively.

Note that for MNPs, besides translational degrees of freedom, the rotational ones matter as well, since the magnetic moment orientations enter the interparticle forces, see Equation (3). The pertinent equations of rotary dynamics with the corresponding torques induced both by the local fields and external field $\vec{H}_0$ are included in the whole set of equation as well.

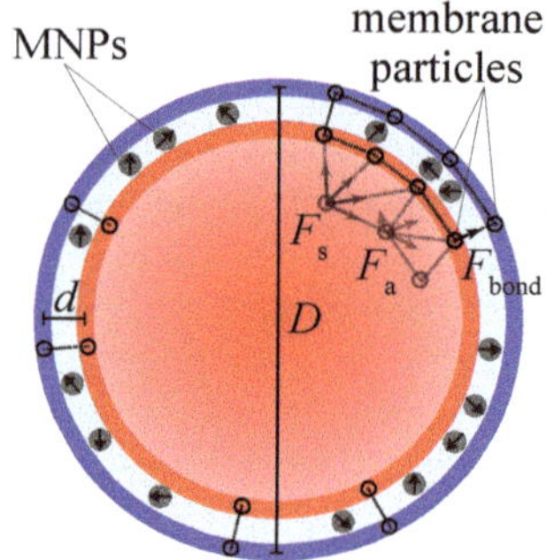

Figure 1. Schematic cross-section of the model MPS. The membrane consists of the internal (red) and external (blue) shells made of equal number of polymer particles (beads) arranged in triangular mesh; the intershell space is filled with MNPs(gray spheres with arrows).

All computer simulations were performed with the aid of ESPResSo code [19], the shells were implemented following the scheme proposed in [20]. The computational experiments included preliminary steps to obtain equilibrium distribution of MNPs inside the MPS membrane under zero external field (basic state) and magnetization steps to obtain the changes induced by the applied field. For every MPS size and for every value of the field strength, the calculations were carried out on 10 copies of the model MPS which differed only in initial orientational distribution of the MNP magnetic moments.

The nondimensional parameters which characterize the system under study are as follows. The intensity of the magnetic dipolar interaction

$$\lambda = \mu_0 \mu^2 / d_p^3 k_B T, \tag{4}$$

is defined as the ratio of energy of magnetic dipolar interaction of two identical MNPs placed at the closest possible distance, i.e., d_p, to thermal energy with k_B being the Boltzmann constant and T temperature.

The second parameter is the reference Zeeman energy of an MNP in the field of strength H_0 scaled with thermal one:

$$\xi = \mu_0 \mu H_0 / k_B T. \tag{5}$$

Let us clarify the relations between the dimensional parameters of the problem and their nondimensional counterparts used in the calculations. The nanoparticle diameter d_p is set to 15 nm that with the typical magnetization of a ferrite (magnetite or maghemite) $M = 500$ emu yields for the MNP magnetic moment $\mu = (\pi/6) M d_p^3 \approx 8 \times 10^{-16}$ emu. Then, assuming room temperature that is the only choice for aqueous solutions, for the dipole interaction parameter (4) one gets $\lambda \approx 5$.

The Langevin argument ξ evaluated for the same conditions comes out is related to dimensional field strength as $\xi \approx 200 H_0$, so that $\xi = 10$ corresponds to $H_0 = 2$ kOe that is completely feasible value.

The intershell distance h is taken as the length unit; in this scale, $d_p/h = 0.3$ and $h_{gap}/h = 0.35$. With the above-introduced value of the MNP diameter $d_p = 15$ nm, for the MPS parameters one finds: $h \approx 50$ nm and $h_{gap} \approx 17$ nm. The size of simulated MPSs varies from $q \equiv D/h = 6$ to $q = 14$ that makes 2.3 time difference; the value of D refers to the initial geometry of the MPS, i.e., a sphere. Taking the above-given value of h, one finds that the dimensional size of the tested MPS spans from 300 to 700 nm. In the membranes of all the samples, the volume content of MNPs is constant and equal $\phi \approx 11$ vol.%, and the connectivity of the MPS shells is set to $c_b = 0.2$.

Keeping the MNP size d_p, dipole parameter λ, layer thickness h_{gap}, volume content ϕ and connectivity c_b constant, we ensure that the "quality" of the MPS membranes is the same (at least, before the field is applied). The next section presents the results of simulating the field-induced responses of such MPSs under variation of their overall sizes D.

3. Results

The tested MPSs are shown schematically in Figure 2 just for visual comparison.

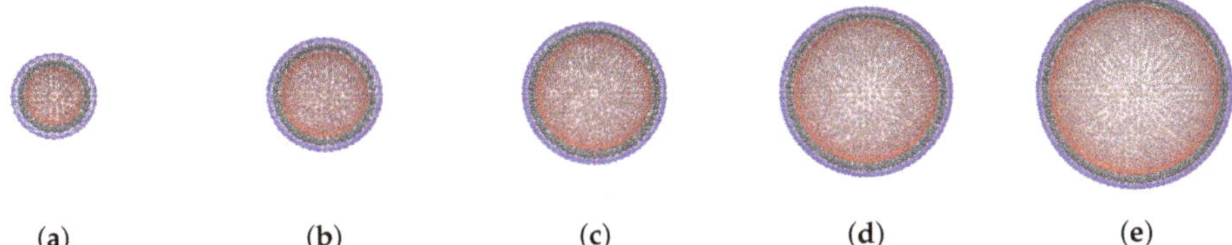

(a)   (b)   (c)   (d)   (e)

Figure 2. Schematic sketches (drawn at the scale) of model MPSs with nondimensional sizes q: 6 (**a**); 8 (**b**); 10 (**c**); 12 (**d**); 14 (**e**); for all samples $\phi \approx 11\%$ and $c_b = 0.2$.

A series of snapshots of equilibrium states for MPSs of different sizes under variation of applied field is presented in Figure 3.

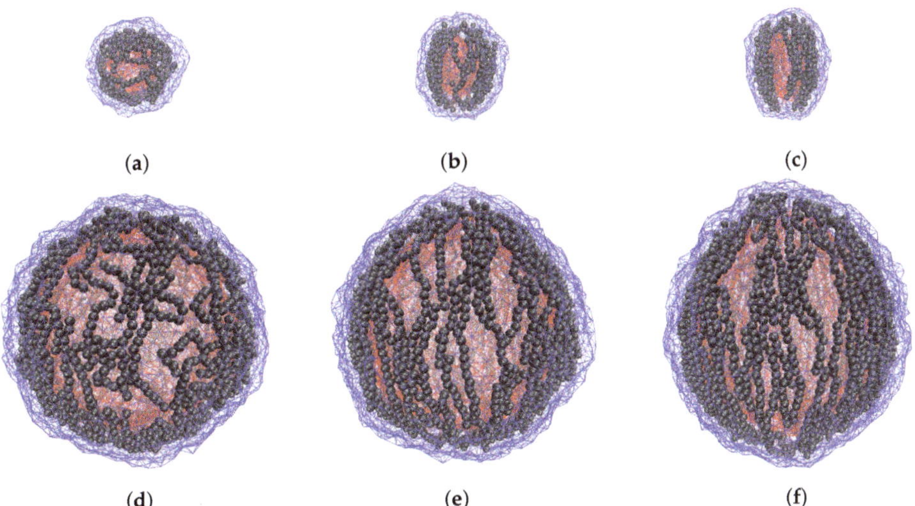

(a)   (b)   (c)

(d)   (e)   (f)

Figure 3. Snapshots of model polymersomes with the size $q = 6$ (**a**)–(**c**) and $q = 14$ (**d**)–(**f**) under applied field $\zeta = 2$ ((**a**) and (**d**)), $\zeta = 6$ ((**b**) and (**e**)), $\zeta = 10$ ((**c**) and (**f**)). The field is directed upward; for all samples $\phi \approx 11\%$ and $c_b = 0.2$.

Simulations of the model MPSs under the field point out the common tendency to stretch along the field direction assuming axisymmetric ellipsoidal-like shape. Because of that, it suffices to describe this deformation by a single elongation parameter ε that we define as the ratio of the major semi-axis of the deformed sample to its minor one minus unity. These results are shown in Figure 4.

The presented plots reveal that the smaller polymersome is subjected to magnetization, the greater is its field-induced deformation. Indeed, under field $\zeta = 10$ the elongation parameter is $\varepsilon \approx 0.25$ for an MPS of size $q = 6$ whereas for the largest one ($q = 14$) it is $\varepsilon \approx 0.17$, i.e., about 30% lower.

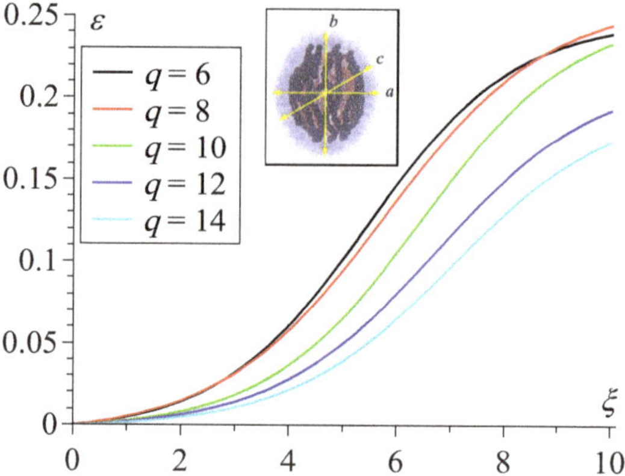

Figure 4. Elongation parameter $\varepsilon = \frac{2b}{(a+c)} - 1$ as a function of external field strength parameter ξ for the MPSs of different initial sizes q; all the curves are obtained for $\phi \approx 11\%$ and $c_b = 0.2$.

Being soft hollow objects, MPSs do not conserve neither their overall nor inner volume under deformation. In view of the drug delivery application, it is interesting to consider the field-induced volume changes of the MPS. The estimation is obtained from simulations by taking any actual configuration of the polymersome and building up a nested polyhedron whose vertexes are fixed at the the centers of the beads of the inner shell. These results are shown in Figure 5, where V_0 corresponds to the basic state of the MPS and is approximately equal to the volume of a sphere of diameter $D - 2h$. As seen, small polymersomes demonstrate a non-monotonic dependence of the "volume defect" $(V(\xi) - V_0)/V_0$ on the field strength whereas larger ones show regular decreasing behavior.

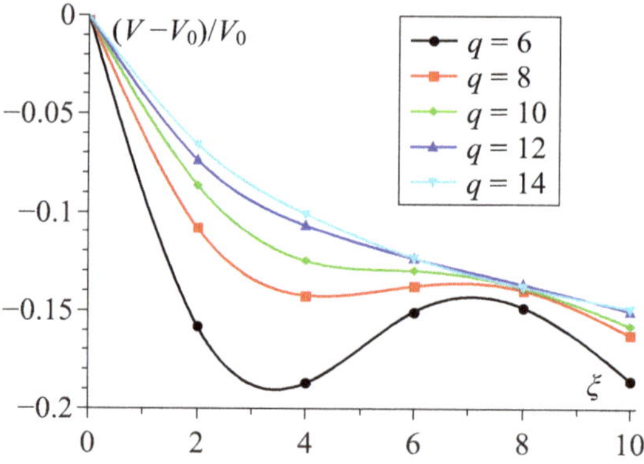

Figure 5. Dependence of the "volume defect" $(V - V_0)/V_0$ of the MPS cavity on applied magnetic field ξ; for all samples $\phi \approx 11\%$ and $c_b = 0.2$.

4. Discussion

The coarse-grained molecular dynamics model is applied to investigate the field-induced overall deformation of an MPS and the internal restructuring accompanying it. This enables one to understand the dimension effect: the influence of the size factor on the polymersomes with the same properties of their membranes. The obtained results point out that in these objects the overall changes (shape anisotropy, volume defect) are essentially determined by the structure rearrangements which take place in the magnetically active membrane. As the snapshots of Figure 3 show, the MNPs, formerly dwelling in randomly oriented loose aggregates, re-group and unite in well-formed chains fairly well aligned with the direction of the imposed field; the increase of the field enhances the straightness of the chains. Therefore, one comes to a conclusion that the field-induced chaining of the particles is the main mechanism of MPS elongation. Moreover, it is clear that the equilibrium shape of an MPS is established as a result of interplay between the magnetostatic forces which group the particles and align the chains, and the steric forces which maintain the MNPs confinement in the intra-membrane space.

At a first sight, in the membrane of a larger polymersome (lower curvature), the nanoparticles are more free to align, and, thus, in such an object the conditions for the field-induced elongation are more favorable. However, as the simulation results presented in Figure 4 show, our modelling points out the opposite tendency. To explain this, in below we present qualitative considerations which, if not being exhaustive, should be an important part of the effect.

In our view, the found dependence is a manifestation of the anisotropic character of the dipole-dipole interaction in a confined geometry of a spherical layer. As known, the dipolar particles with parallel orientation of their magnetic moments attract each other if positioned in a "head-to-tail" pattern and experience mutual repulsion in "side-by-side" configuration. More broadly, the particles repel each other if their center-to-center vector is inclined to the field under the angles $\Delta\psi \approx (90 \pm 35)°$ and are mutually attracted in all the other positions.

When such particles are enclosed in a spherical layer, which is the case of an MPS subjected to a strong field, the situations in the "equatorial" and "polar" zones of the membrane are qualitatively different. (For convenience, we liken an MPS to a globe whose poles are at the points where the direction of external field is parallel to the surface normal.) Indeed, as long as the MPS is sufficiently large $D \gg d_{\mathrm{p}}$, those MNPs that inhabit the equatorial zone, are in quite favorable situation for building meridional chains of the "head-to-tail" structure . Meanwhile, for those which at zero field occupied the near-polar zone, the situation strongly depends on the curvature of the layer.

At high curvature (small MPS), virtually all feasible center-to-center vectors fall outside the $\Delta\psi$ interval. This implies that the formerly "polar" particles are attracted by the ends of the meridional chains, drift in their directions, and, thus, deplete the polar zone. At low curvature (large MPS), for a notable number of particles in the near-polar zone have their center-to-center vectors inside the $\Delta\psi$ interval and, thus, experience mutual repulsion. Instead of joining long meridional chains, those MNPs would be repelled by their ends and would stably dwell in the polar zone as single entities or very short weakly linked aggregates. An illustration for this conclusion is given by snapshots in Figure 6.

Evidently, those non-chained particles do not contribute to the stretching effect and by that reduce to some extent the number and/or length of the formed chains. Moreover, the repulsion exerted by those "polar" particles on the nearly-situated chains should extend the MPS in the direction transverse to the field direction and, albeit weakly, but also contribute to diminution of the field-induced stretching of larger MPSs. In Appendix A we propose a "toy" model that justifies the effect of this chain-length reduction on the size-dependent magnetic deformation of MPSs.

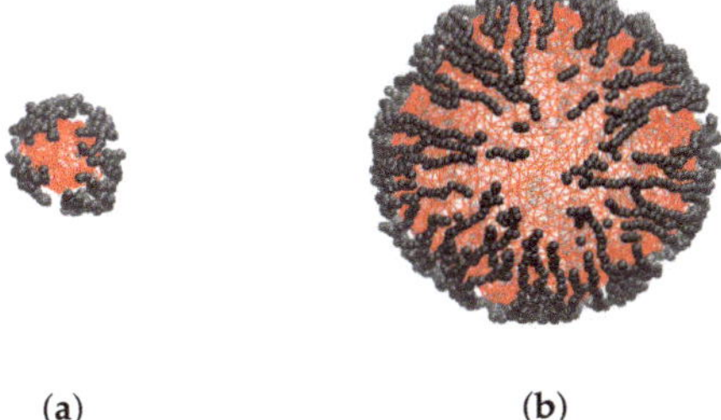

(a)   **(b)**

Figure 6. Top views of the MPSs with sizes $q = 6$ (**a**) and 14 (**b**) magnetized by field $\xi = 10$; to improve vision, the outer shell is not shown; these top views correspond to the side views given in Figure 3c,f.

Another effect that is worth of discussion is the non-monotonic field dependence of the inner volume of small polymersomes, see Figure 5. In general, as it is could be shown analytically, when a sphere is deformed to spheroid under condition of constancy of its lateral area, the "volume defect" is always negative. Therefore, the found negative volume change, see Figure 5, confirms the assumption of high surface tension of the membrane shells. Looking to details, we surmise that the initial stretching of the MPS in the low field range ($\xi = 0$ to ≈ 3), see Figure 3 (a), where the chains are very curvilinear, is due to orientation of some chain fragments without breaking their overall structure arrangement.

In small polymersomes, on further growth of the field ($\xi \approx 4$ to ≈ 8), the structure rearrangement requires numerous breaks of the formerly existing chains. When the extensive chain rupture takes place, the particle structure becomes more loose, and this transformation, although not completely (see curve for $q = 6$ in Figure 5), but to some extent reduces the volume defect. As soon as the chain fragments unite in a new conformations which are plausible for making quasi-meridional lines, the diminution of V resumes. Finally, the field-induced restructuring results in building up inside the membrane a barrel-like superstructure (we remind of the afore-mentioned effect of depleting the polar zones) that makes a small MPS to efficiently stretch along the field direction, see Figure 3c. In large polymersomes, the already existing chains have much more freedom for their conformational changes and perform them with almost no breaks. Therefore the initially formed aggregates gradually reorganize in long-chain patterns; because of that, in such an MPS compression of the inner cavity goes in monotonic way, see plots for $q = 12$ and $q = 14$ in Figure 5 and Figure 3d–f.

5. Conclusions

The size effect, i.e., the dependence of structure and deformation of a magnetic polymersome on the applied field strength is studied with the aid of a coarse-grained model. It is shown that in the realistic size range, the field-induced strain—the polymersomes elongation along the applied field—is the higher the smaller the objects. This tendency is related to the arrangement and size of the nanoparticle chains which form and align in the polymersome membrane under magnetization. The field-induced change of the polymersome inner cavity volume is negative, and for small polymersomes its field dependence turns out to be non-monotonic.

The presented results infer that in terms of specific effects—the field-induced strain as well as the volume defect—smaller polymersomes turn out to be more efficient than the larger ones. From application view point, the volume defect is important, in the first place, for drug delivery. The smallness of the obtained value indicates that polymersomes of the considered kind hardly have wide prospects. On the other hand, the 20–25% strain effect might fit in a fairly good way the requirements for cell membrane stimulations which are realized either via turning an already stretched polymersome by a rotating field or by inducing its periodic elongation in the given direction with the aid of alternating field. Certainly, the obtained results on deformational properties of magnetic polymersomes, apart from static cases, might also provide reliable estimates for sufficiently low-frequency magnetodynamic regimes.

Besides the above-presented particular points, it is worth noting that the calculation model that we use, may be easily modified and/or extended in many ways that is an advantage if one would need to describe a magnetic polymersome in more detail.

Author Contributions: Conceptualization, A.R. and Y.R.; data curation, A.R.; formal analysis, A.R.; funding acquisition, A.R. and Y.R.; investigation A.R. and Y.R.; methodology A.R. and Y.R.; project administration, Y.R.; resources (computing resources), ICMM UB RAS; software, A.R.; supervision, Y.R.; visualization, A.R.; writing–original draft preparation, A.R. and Y.R.

Funding: The research was funded by Russian Foundation for Basic Research under grant #17-42-590504.

Acknowledgments: The work was performed using *Triton* supercomputer of ICMM UB RAS. Authors are indebted to Olivier Sandre for valuable discussions.

Conflicts of Interest: The authors declare no conflict of interest. The funders had no role in the design of the study; in the collection, analyses, or interpretation of data; in the writing of the manuscript, or in the decision to publish the results.

Abbreviations

The following abbreviations are used in this manuscript:

MNP	magnetic nanoparticle
MPS	magnetic polymersome
MRI	magnetic resonance imaging
PS-*b*-PAA	poly(styrene)-*b*-poly(acrylic acid)
PTMC-*b*-PGA	poly(trimethylene carbonate)-*b*-poly(L-glutamic acid)
SANS	small angle neutron scattering

Appendix A

Consider an assembly of identical single-domain spherical particles coupled with each other via a sufficiently strong dipolar potential. Under an imposed external field, the basic state (lowest energy) configuration of such a system is a perfectly straight line parallel to the field direction. Evidently, any distortion of this contour enhances the energy of the chain and generates the forces striving to recover the basic state. If the chain is confined, these forces are exerted on the boundaries/walls which restrict its conformational freedom. This is the case of a magnetized MPS: in a strong external field the MNPs align their magnetic moments with the field and aggregate in chains but, being inside the membrane, those chains come out arc-shaped with the curvature equal to that of the membrane in the "meridional" plane, i.e., the plane that contains the field vector and passes through the MPS center. The tendency of the confined chains to straighten, imparts to the membrane an excess of surface energy, and this energy diminishes as the MPS stretches in the field direction.

In an MPS, the approximate volume of spherical layer accessible for MNPs is $V = \pi D^2 h_{gap}$, whereas the total volume of all N particles is $(\pi/6)d_p^3 N$. Taking into account that the intershell gap is close to the MNP diameter ($h_{gap} \approx d_p$), the volume fraction of MNPs comes out as $\phi = N d_p^2 / 6D^2$. We remind that the above-presented modelling is performed under condition $\phi = $ const.

Let us assume that the chains stretch meridionally and they are symmetrical with respect to equator of the membrane, so that the chain length is νd_p; the maximal value of the length is $\nu_0 = \pi D/2d_p$, i.e., the chain makes a pole-to-pole semi-circle. The expression for energy ΔU for a model chain of constant curvature could be obtained from general expressions for long dipolar chains, see [21], for example, and taken in the form

$$\Delta U \approx 6\zeta(3)\left(\nu\mu^2/d_p^3\right)\left[1 - \cos^2\vartheta\right]. \tag{A1}$$

As the chain is considered to be long enough (say, $\nu > 10$), it is described in continual limit; here $\zeta(3)$ is Riemann's function and ϑ the local angle between the tangent to the contour of the chain and the field direction; note that we count ΔU from the lowest energy, where $\cos^2\vartheta = 1$. Under assumption of constant curvature of the arc $(2/D)$, the cosine function in expression (A1) could be averaged over the chain and yields

$$\cos^2\vartheta = \frac{1}{2}\left[1 + \frac{\sin(\pi\nu/\nu_0)}{\pi\nu/\nu_0}\right], \tag{A2}$$

so that substitution to (A1) gives

$$\Delta U \approx 3\zeta(3)\frac{\nu\mu^2}{d_p^3}\left[1 - \frac{\sin(\pi\nu/\nu_0)}{\pi\nu/\nu_0}\right]. \tag{A3}$$

The part of the membrane surface per one chain is $\Delta S = \pi D^2 v / N = \pi v d_{\rm p}^2 / 24\phi$. Then the surface energy density in question is

$$\sigma = \Delta U / \Delta S = 72\zeta(3)\frac{\phi\mu^2}{d_{\rm p}^5}\left[1 - \frac{\sin(\pi v/v_0)}{\pi v/v_0}\right]. \tag{A4}$$

As seen, in expression (A4) the coefficient before the bracket is independent of the MPS structure. Had the ratio v/v_0 been the same for any of the considered MPSs, then this independence would have been inherent to the bracket as well. However, as pointed out in section 4, the larger the MPS diameter D, the greater number of MNPs does not take part in stretching. This means that in larger MPSs the ratio v/v_0 is lower than in smaller ones. This entails diminution of the value of the factor in brackets and, thus, the decrease of the excess of the magnetic surface energy σ that, in our view, is the driving mechanism of the field-induced stretching of MPSs. To illustrate this, the v-dependent function of Equation (A4) is presented in Figure A1.

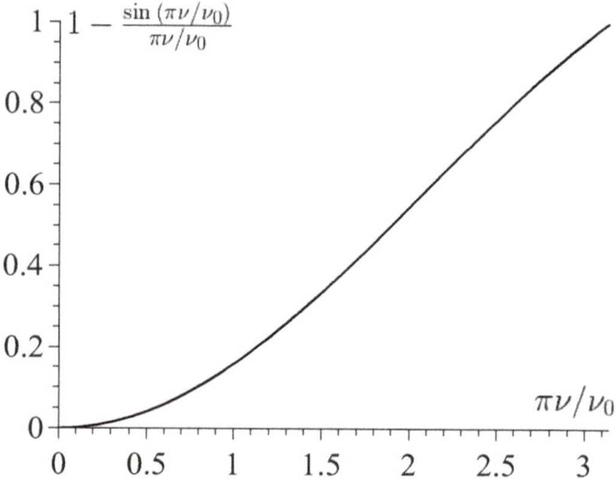

Figure A1. The factor depending on the relative chain length in Equation (A4).

References

1. Letchford, K.; Burt, H. A review of the formation and classification of amphiphilic block copolymer nanoparticulate structures: Micelles, nanospheres, nanocapsules and polymersomes. *Eur. J. Pharm. Biopharm.* **2007**, *65*, 259–269. [CrossRef] [PubMed]
2. Palivan, C.G.; Goers, R.; Najer, A.; Zhang, X.; Car, A.; Meier, W. Bioinspired polymer vesicles and membranes for biological and medical applications. *Chem. Soc. Rev.* **2016**, *45*, 377–411. [CrossRef] [PubMed]
3. Che, H.; van Hest, J.C.M. Stimuli-responsive polymersomes and nanoreactors. *J. Mater. Chem. B* **2016**, *4*, 4632–4647. [CrossRef]
4. Balasubramanian, V.; Herranz-Blanco, B.; Almeida, P.V.; Hirvonen, J.; Santos, H.A. Multifaceted polymersome platforms: Spanning from self-assembly to drug delivery and protocells. *Prog. Polym. Sci.* **2016**, *60*, 51–85. [CrossRef]
5. Mohammadi, M.; Ramezani, M.; Abnous, K.; Alibolandi, M. Biocompatible polymersomes-based cancer theranostics: Towards multifunctional nanomedicine. *Int. J. Pharm.* **2017**, *519*, 287–303. [CrossRef] [PubMed]
6. Zhu, Y.; Yang, B.; Chen, S.; Du, J. Polymer vesicles: Mechanism, preparation, application, and responsive behavior. *Prog. Polym. Sci.* **2017**, *64*, 1–22. [CrossRef]
7. Hu, X.; Zhang, Y.; Xie, Z.; Jing, X.; Bellotti, A.; Gu, Z. Stimuli-responsive polymersomes for biomedical applications. *Biomacromolecules* **2017**, *18*, 649–673. [CrossRef]
8. Oliveira, H.; Pérez-Andrés, E.; Thevenot, J.; Sandre, O.; Berra, E.; Lecommandoux, S. Magnetic field triggered drug release from polymersomes for cancer therapeutics. *J. Controll. Release* **2013**, *169*, 165–170. [CrossRef] [PubMed]

9. Bleul, R.; Thiermann, R.; Marten, G.U.; House, M.J.; Pierre, T.G.; Häfeli, U.O.; Maskos, M. Continuously manufactured magnetic polymersomes—A versatile tool (not only) for targeted cancer therapy. *Nanoscale* **2013**, *5*, 11385. [CrossRef] [PubMed]

10. Liu, Q.; Song, L.; Chen, S.; Gao, J.; Zhao, P.; Du, J. A superparamagnetic polymersome with extremely high T2 relaxivity for MRI and cancer-targeted drug delivery. *Biomaterials* **2017**, *114*, 23–33. [CrossRef] [PubMed]

11. Geilich, B.M.; Gelfat, I.; Sridhar, S.; van de Ven, A.L.; Webster, T.J. Superparamagnetic iron oxide-encapsulating polymersome nanocarriers for biofilm eradication. *Biomaterials* **2017**, *119*, 78–85. [CrossRef] [PubMed]

12. Salva, R.; Le Meins, J.F.; Sandre, O.; Brûlet, A.; Schmutz, M.; Guenoun, P.; Lecommandoux, S. Polymersome shape transformation at the nanoscale. *ACS Nano* **2013**, *7*, 9298–9311. [CrossRef] [PubMed]

13. Rikken, R.S.M.; Engelkamp, H.; Nolte, R.J.M.; Maan, J.C.; van Hest, J.C.M.; Wilson, D.A.; Christianen, P.C.M. Shaping polymersomes into predictable morphologies via out-of-equilibrium self-assembly. *Nat. Commun.* **2016**, *7*, 12606. [CrossRef] [PubMed]

14. Sanson, C.; Diou, O.; Thévenot, J.; Ibarboure, E.; Soum, A.; Brûlet, A.; Miraux, S.; Thiaudière, E.; Tan, S.; Brisson, A.; et al. Doxorubicin loaded magnetic polymersomes: Theranostic nanocarriers for MR imaging and magneto-chemotherapy. *ACS Nano* **2011**, *5*, 1122–1140. [CrossRef] [PubMed]

15. Yang, K.; Liu, Y.; Liu, Y.; Zhang, Q.; Kong, C.; Yi, C.; Zhou, Z.; Wang, Z.; Zhang, G.; Zhang, Y.; Khashab, N.M.; et al. Cooperative assembly of magneto-nanovesicles with tunable wall thickness and permeability for MRI-guided drug delivery. *J. Am. Chem. Soc.* **2018**, *140*, 4666–4677. [CrossRef] [PubMed]

16. Hickey, R.J.; Koski, J.; Meng, X.; Riggleman, R.A.; Zhang, P.; Park, S.J. Size-controlled self-assembly of superparamagnetic polymersomes. *ACS Nano* **2014**, *8*, 495–502. [CrossRef] [PubMed]

17. Ryzhkov, A.; Raikher, Y. Coarse-grained molecular dynamics modelling of a magnetic polymersome. *Nanomaterials* **2018**, *8*, 763. [CrossRef] [PubMed]

18. Weeks, J.D.; Chandler, D.; Andersen, H.C. Role of repulsive forces in determining the equilibrium structure of simple liquids. *J. Chem. Phys.* **1971**, *54*, 5237–5247. [CrossRef]

19. Arnold, A.; Lenz, O.; Kesselheim, S.; Weeber, R.; Fahrenberger, F.; Roehm, D.; Košovan, P.; Holm, C. ESPResSo 3.1: Molecular dynamics software for coarse-grained models. In *Lecture Notes in Computational Science and Engineering*; Griebel, M., Schweitzer, M.A., Eds.; Springer: Berlin/Heidelberg, Germany, 2013; Volume 89, pp. 1–23. [CrossRef]

20. Cimrák, I.; Gusenbauer, M.; Jančigová, I. An ESPResSo implementation of elastic objects immersed in a fluid. *Comput. Phys. Commun.* **2014**, *185*, 900–907. [CrossRef]

21. Hall, C.L.; Vella, D.; Goriely, A. The mechanics of a chain or ring of spherical magnets. *SIAM J. Appl. Math.* **2013**, *73*, 2029–2054. [CrossRef]

Article

Modulating the Spin Seebeck Effect in Co$_2$FeAl Heusler Alloy for Sensor Applications

Marcus Vinicius Lopes [1,2], Edycleyson Carlos de Souza [1], João Gustavo Santos [1], João Medeiros de Araujo [1], Lessandro Lima [1], Alexandre Barbosa de Oliveira [1], Felipe Bohn [1] and Marcio Assolin Correa [1,*]

[1] Departamento de Física, Universidade Federal do Rio Grande do Norte, Natal 59078-900, RN, Brazil
[2] Instituto Federal de Educação Ciência e Tecnologia do Ceará, Quixadá 63902-580, CE, Brazil
* Correspondence: marciocorrea@fisica.ufrn.br

Received: 06 February 2020; Accepted: 29 February 2020 ; Published: 3 March 2020

Abstract: The thermoelectric conversion technique has been explored in a broad range of heat-flow sensors. In this context, the Spin Seebeck Effect emerges as an attractive candidate for biosensor applications, not only for the sensibility improvement but also for the power-saving electronic devices development. Here, we investigate the Longitudinal Spin Seebeck Effect in films with a Co$_2$FeAl/W bilayer structure grown onto GaAs (100) substrate, systems having induced uniaxial magnetic anisotropy combined with cubic magnetic anisotropy. From numerical calculations, we address the magnetic behavior and thermoelectric response of the films. By comparing experiment and theory, we explore the possibility of modulating a thermoelectric effect by magnetic anisotropy. We show that the thermoelectric voltage curves may be modulated by the association of magnetic anisotropy induction and experimental parameters employed in the LSSE experiment.

Keywords: spintronics; CFA; thermoelectric effect; spin seebeck effect

---

1. Introduction

Presently, the thermoelectric effects, such as the Seebeck effect, are widely explored in the direct conversion of heat-flow in voltage signal [1–3]. However, usually, these systems present a high heat resistance, leading to limitations in biosensor applications [4]. On the other hand, thermoelectric effects based on spin dynamics have received increasing attention in recent past, not just to the great technological potential in power-saving electronic devices [5–9], but also for photodetectors [10,11], diode [12], and temperature sensor [13].

As a result, numerous studies have been performed in a broad range of nanostructures, promoting significant advances in the performance of such effects [4,14,15].

Within this field, Anomalous Nernst (ANE) and the Longitudinal Spin Seebeck (LSSE) effects have distinguished positions, emerging as interesting phenomena for biosensors and spin caloritronics devices [16–18]. Both effects generally consist in the application of a temperature gradient ∇T and a magnetic field $\vec{H}$, thus generating an electric field $\vec{E}$ that can be expressed as [19]

$$\vec{E} = -S(\hat{m} \times \nabla T), \tag{1}$$

where $S = \lambda \mu_{\circ} m_s$, $\mu_{\circ}$ is the vacuum magnetic permeability, and m_s is the saturation magnetization of the ferromagnetic alloy, which is oriented along to the unit vector $\hat{m}$. Here, λ is a fundamental coefficient that is straightly related to the nature of the involved effect. Specifically, it is common to assume $\lambda = \lambda_N$, i.e., the well-known Anomalous Nernst coefficient, for a sample consisting of a single ferromagnetic metallic layer in which ANE is the unique effect in the system. However, for a bilayer structure composed by a ferromagnetic metallic layer capped by a non-magnetic metal in turn, an effective λ_{eff} coefficient shall be taken into consideration. Remarkably, for this latter, both ANE and LSSE contribute to the generation of the electric field $\vec{E}$.

In a typical LSSE experiment, the electric field associated with the effect is measured through the Inverse Spin Hall Effect (ISHE). Nevertheless, we may find in the literature several studies uncovering distinct ways to overcome this experimental adversity, split the contributions of the LSSE and ANE effects [5,20], or improve the thermoelectric conversion efficiency. For instance, very recently, Holanda and colleagues [20] proposed a Si/NiFe/NiO/Pt heterostructure to detect the ANE and LSSE thermoelectric voltages simultaneously, thus allowing the comparison of the results with the values acquired for a single Si/NiFe film. Following a distinct approach, aiming to obtain optimized responses, one may focus on bilayer structures and play with the choice of the ferromagnetic material and the metallic capping layer, both having deep impacts on the thermoelectric conversion efficiency. In this case, it is well known that capping-layer materials with high spin-orbit coupling are the main elected for investigations of spintronics effects [21]. This fact is justified given that they enable the interconversion between charge current and spin current [5]. Nonetheless, the ferromagnetic material has a fundamental role in this interconversion, being responsible for the spin current injection. Therefore, properties as magnetic permeability, magnetic anisotropy, damping parameter, and capability to generate spin-polarized current are key features that must be tuned to improve the thermoelectric effects in bilayer structures [22–24].

Within this spirit, Heusler alloys appear as attractive candidates for thermoelectric applications [25–27], given that in theory, they can present energy gap around the Fermi level, making possible the achievement of 100% spin-polarized currents [28]. Among the numerous Heusler alloys, Co_2FeAl (CFA) is a conspicuous material due to its high magnetic permeability and controllable magnetic anisotropies [29,30], features often found in CFA films. Theoretically, this alloy has been widely studied through first-principle calculations [31–34]. Experimentally, the magnetic properties in CFA films are strongly dependent of parameters used during the production process, such as temperature deposition and annealing after the deposition [35], as well as on the employed substrate. Therefore, the control of the features of the CFA alloy as magnetic anisotropy [35–38], associated with the integration between electrical and magnetic properties, may open new roads to explore this material and to modulate its thermoelectric response.

In this work, we explore the possibility of modulating a thermoelectric effect by magnetic anisotropy. Specifically, we investigate the Longitudinal Spin Seebeck Effect in films with a CFA/W bilayer structure grown onto GaAs (100) substrate, systems with induced uniaxial magnetic anisotropy combined with cubic magnetic anisotropy. From numerical calculations, we address the magnetic behavior and thermoelectric response of the films. By comparing experiment and theory, we show that the thermoelectric voltage curves may be controlled by the association of magnetic anisotropy induction and experimental parameters employed in the LSSE experiment. These results enable us to modulate the thermoelectric response as a function of the applied magnetic field, opening new possibilities for sensor application.

2. Materials and Methods

Here, we perform experiments in a set of films with a CFA/W bilayer structure, in which the CFA layer in each sample was grown at a given temperature. Specifically, CFA layers were prepared at the selected temperatures of 300, 573, 673 and 773 K [35]. The thickness of the ferromagnetic and metallic non-magnetic layers in the bilayer structure is 53 and 2 nm, respectively. The films were deposited by magnetron sputtering onto GaAs (100) substrates with dimensions of 3×6 mm^2, previously annealed at 923 K, and covered by a 2-nm-thick W buffer layer. The deposition process was carried out with the following parameters: base pressure of 7.8×10^{-8} Torr, Ar pressure during disposition of 3×10^{-3} Torr with flow of 14 sccm, current of 200 mA set in the DC source for the CFA layer deposition, and 140 mA set in similar source for the grown of the W layer. Using these parameters, we reach deposition rates were 0.6 nm/s and 0.53 nm/s for the CFA and W, respectively. During the whole production process for all samples, a constant 1.0 kOe magnetic field was applied perpendicularly to the main axis of the substrate.

The structural characterization of the samples was obtained by x-ray diffraction. The experiments were performed using CuK_α radiation. While low-angle x-ray diffraction was employed to determine the deposition rate and calibrate the film thickness, high-angle x-ray diffraction measurements were used to verify the structural character of the samples.

Quasi-static magnetic characterization was obtained using a LakeShore Model 7404 Vibrating Sample Magnetometer (VSM). Specifically, the in-plane magnetic properties were verified through magnetization curves, acquired with a maximum magnetic field value of ± 500 Oe. The set up allowed us the rotation of the sample. Thus, to verify the magnetic anisotropy induced in the films, we measured magnetization curves for the different φ values, the angle between the magnetic field and the shortest axis of the samples.

Ferromagnetic resonance (FMR) measurements were carried out with a Bruker EPR system operating at 9.838 GHz (TE011 mode). In this case, the sample was placed in the center of a cylindrical cavity, in which the microwave magnetic field is maximized, and the electrical field is minimized. The set up also allowed us the rotation of the sample and experiments were performed at different φ values. The magnetic field, applied in the film plane, varied from 2.0 up to 3.5 kOe. It was modulated in the amplitude of 0.1 Oe at 100 kHz. Due to the modulation, the FMR power absorption curve took the shape of an absorption derivative, which was fitted by a Lorentzian-derivative function to extract resonance field and linewidth.

At last, we used a home-made Longitudinal Spin Seebeck Effect system (LSSE) to perform thermoelectric voltage measurements. Further information on the experimental system can be found in Ref. [39]. The system consists of a heat source (Peltier modulus) and a heat sink composed by a Cu block, in which the film is placed between them. The thermal conductivity was improved by the thermal paste, while the electrical contacts were made using silver paint. The system (heat sources and film) was connected to a step-motor, making possible the sample rotation and the measurement of the thermoelectric voltage at different φ values. Here, following the previous definition considered for the magnetic characterization, we kept the definition of φ as the angle between the magnetic field and the shortest axis of the sample. Figure 1 depicts a schematic representation of our bilayer structure and the LSSE experiment.

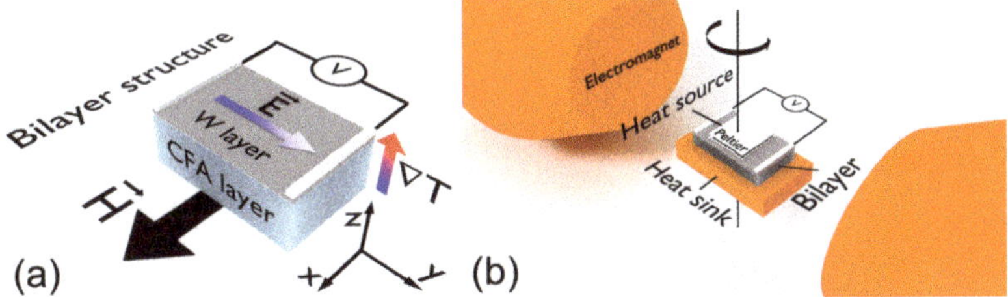

Figure 1. Schematic representation of our bilayer structure and the LSSE experiment. (**a**) Bilayer structure and definition of the coordinate system used in the numerical calculations. Given that φ is the angle between the magnetic field and the shortest axis of the sample, $\varphi = 0°$ is found when $\vec{H}$ is along the x axis. (**b**) LSSE experimental setup employed for the thermoelectric voltage measurements. The system allowed us the rotation of the sample during the experiment. Notice that the configuration represented in (**b**) corresponds to $\varphi = 90°$.

3. Results and Discussion

It is known that the structural features of films have a strong influence on their magnetic properties.

In ferromagnetic films produced onto amorphous substrates, a considerable out-of-plane anisotropy contribution is observed as the thickness is increased [40]. This behavior is related to the local stress storage in the film. On the other hand, by employing oriented substrates, we may promote a significant reduction of the local stored stress, favoring the appearance of magnetic anisotropy of magnetocrystalline nature.

This is precisely our intent in this work, in which we made use of an oriented substrate and annealing during the deposition to produce films having induced uniaxial magnetic anisotropy combined with cubic magnetic anisotropy.

3.1. Structural Results

Regarding the structural properties, Figure 2 shows the high-angle XRD results obtained for our set of films with CFA/W bilayer structure grown onto GaAs (100) substrate. First, the diffractograms disclose the (002) planes of the GaAs substrate [41–43], assigned by the well-defined and high-intensity peak located at $2\theta \approx 31.9°$. Second, no evidence of the W layer is verified here, as expected, a fact that is associated with the reduced thickness of this layer. Going beyond, with respect to the CFA layer, the results clearly reveal a peak located at $2\theta \approx 44.73°$ that is a fingerprint of the (022) texture of the CFA alloy, and peaks at $2\theta \approx 31.55°$ and $59.15°$, which are signatures of the CFA (002) texture. All these CFA peaks are identified from the pattern found in ICDD 01-0715670. Similar structural results have been previously reported in the literature for CFA films grown onto distinct oriented substrates [35,37]. In particular, the peaks observed here enable us to infer the polycrystalline character of the CFA alloy, with an A2 structure, suggesting disorder of the Co, Fe, and Al sites [38,44,45].

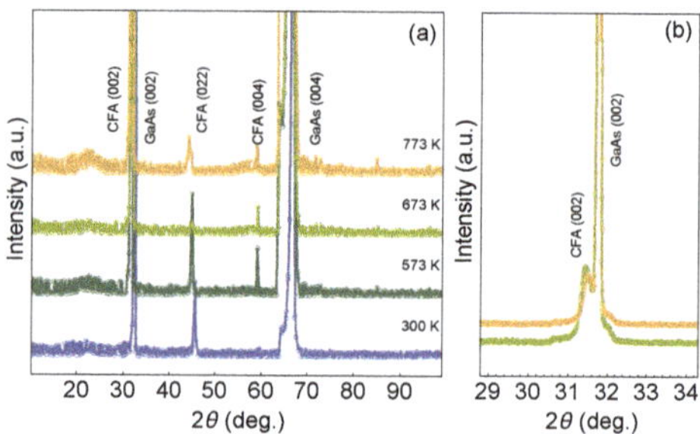

Figure 2. (**a**) High-angle x-ray diffraction results for the Co$_2$FeAl/W bilayers grown onto GaAs (100) substrate. (**b**) Detailed view of the curves, in which the CFA(002) peak can be visualized. The CFA pattern is obtained from ICDD 01-0715670. Notice the polycrystalline character of the CFA alloy in all samples of the set.

3.2. Magnetic Properties

With respect to the magnetic properties, Figure 3 presents the magnetization curves measured for the CFA/W bilayer films at selected φ values. The angular dependence of the magnetization curves confirms the induction of magnetic anisotropy in all films. Notice the remarkable differences among the curves obtained at different φ values, especially for the samples produced with the temperatures of 573 K and 673 K. For all studied bilayers, the curve acquired at $\varphi = 0°$ suggests the existence of an intermediate magnetic axis along this direction. Furthermore, the results reveal an easy magnetization axis along the direction of $\varphi = 90°$, as well as a hard axis close to $\varphi = 45°$ (This latter is clearly observed through the FMR measurements, as we will see in the following). At last, no evidence of an out-of-plane anisotropy contribution is verified in the curves. Our results also corroborate previous studies reported in the literature, in which the CFA films were grown onto oriented substrates, such as Si, MgO, and GaAs [16,17,35]. All these features uncover the combination of induced uniaxial magnetic anisotropy and cubic magnetic anisotropy in the bilayers [46].

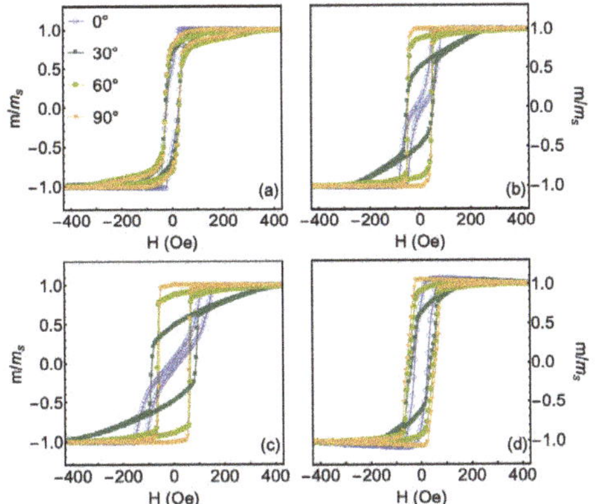

Figure 3. Normalized magnetization curves for the Co$_2$FeAl/W bilayers grown onto GaAs (100) substrate, in which the CFA layer was deposited at the temperatures of (**a**) 300 K, (**b**) 573 K, (**c**) 673 K, and (**d**) 773 K. Here we present just the curves measured for $\varphi = 0°, 30°, 60°$ and $90°$. All the samples of the set present uniaxial magnetic anisotropy combined with cubic magnetic anisotropy, although these features are more evident for the ones produced at the temperatures of 573 K and 673 K.

3.3. Modeling a System with Uniaxial Magnetic Anisotropy Combined with Cubic Magnetic Anisotropy

Given all the state above, we took into account a modified Stoner–Wohlfarth model [39,47] to mimic our Co$_2$FeAl/W bilayers having induced uniaxial magnetic anisotropy combined with cubic magnetic anisotropy. For the numerical calculations, we considered a system with 50 non-interacting magnetic domains, in which the free energy density for each domain is described by

$$\zeta_i = -\vec{m}_i \cdot \vec{H} + 4\pi m_{si}^2 (\hat{m}_i \cdot \hat{n})^2 - k_{ui}(\hat{m}_i \cdot \hat{u}_{ki})^2 - \frac{1}{4}(\xi_{c1i} + \xi_{c2i}). \tag{2}$$

Here, the first term is associated with the Zeeman interaction, the second one describes the shape anisotropy, the third tells on the uniaxial magnetic anisotropy, and the last term is related with the magnetocrystalline anisotropy. Furthermore, the considered quantities are defined as the following: $\hat{m}_i$ is the unit vector of the magnetization, m_{si} is the saturation magnetization, $\vec{H}$ is the magnetic field, $\hat{n}$ the unit vector normal to the film plane ($\theta_n = 0°$), and $k_{ui} = \frac{m_{si}h_{ki}}{2}$, where h_{ki} is the anisotropy field related to the uniaxial magnetic anisotropy oriented along $\hat{u}_{ki}$ for each domain. At last, for the cubic magnetic anisotropy energy density,

$$\xi_{c1i} = k_{c1i}\left(\alpha_{1i}^2\alpha_{2i}^2 + \alpha_{1i}^2\alpha_{3i}^2 + \alpha_{2i}^2\alpha_{3i}^2\right), \tag{3}$$

and

$$\xi_{c2} = k_{c2}\left(\alpha_{1i}^2\alpha_{2i}^2\alpha_{3i}^2\right) \tag{4}$$

where α_{1i}, α_{2i}, and α_{3i} are the components of the unit vector of the magnetization, $\alpha_{1i} = \cos\varphi_{mi}\sin\theta_{mi}$, $\alpha_{2i} = \sin\varphi_{mi}\sin\theta_{mi}$ and $\alpha_{3i} = \cos\theta_{mi}$.

For the case in which the uniaxial anisotropy lies in the film plane and the magnetic field is also applied plane, Equation (2) is drastically simplified and can be rewritten as

$$\zeta_i = -m_{si}\,H\,\cos(\varphi - \varphi_{mi}) - k_{ui}\,\sin(\varphi_{ki} - \varphi_{mi})^2 - \frac{1}{4}k_{c1i}(\cos(\varphi_{mi})^2\sin(\varphi_{mi})^2). \tag{5}$$

From the minimization of Equation (5) for a given magnetic field $\vec{H}$, we are able to find the equilibrium angle of the magnetization of the $i - th$ domain. This angle is a result of the competition between the contributions of the magnetic field and the magnetic anisotropies in the system. To account the whole magnetic behavior, in turn, we also need to consider the magnetic anisotropy dispersion of the system. Further information on the numerical calculations will be provided in the next sections, combining theory and experiment.

3.4. Magnetic Response at Saturated State

Moving forward, evidence of the combination of anisotropies was also investigated through ferromagnetic resonance experiments. In particular, angular FMR measurements enable us to confirm the observed anisotropies, as well as to infer their relative contribution to the whole magnetic behavior. From the FMR absorption derivative signal obtained in the experiments for our samples, not shown here, we determined the linewidth ΔH and resonance field H_r, this latter presented in Figure 4.

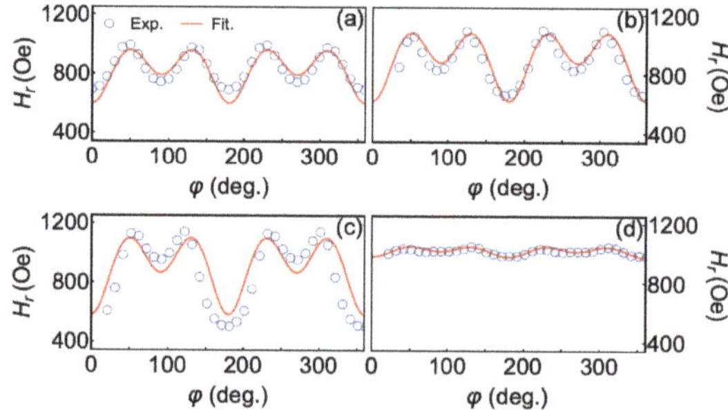

Figure 4. Angular dependence of the resonance field H_r obtained from the FMR absorption derivative signal measured for the Co_2FeAl/W bilayers in which the CFA layer was grown at (**a**) 300 K, (**b**) 573 K, (**c**) 673 K, and (**d**) 773 K. The red lines are fittings obtained with Equations (6) and (5).

From a general point of view, all bilayer films share the very same angular dependence of the resonance field H_r. In particular, the curves are characterized by four peaks with similar amplitude and four valleys, which are split in two groups according to the amplitude values. This behavior is a clear signature of the uniaxial magnetic anisotropy combined with cubic magnetic anisotropy. Furthermore, as expected, we confirm here that the samples produced with the temperatures of 573 K and 673 K do have more intense magnetic anisotropies, which are depicted by the larger amplitude variations of H_r.

In order to obtain further information of our system and estimate anisotropy constants, we considered the well-known angular resonance frequency ω_r and linewidth of the resonance absorption $\Delta\omega$, which can be written as [24,48,49]

$$\omega_r = \frac{\gamma}{m_s \sin\theta_m}\sqrt{1+\alpha^2}\sqrt{\zeta_{\theta\theta}\zeta_{\varphi\varphi} - \zeta_{\theta\varphi}^2}\,,\tag{6}$$

and

$$\Delta\omega = \frac{\alpha\gamma}{m_s}\left(\zeta_{\theta\theta} + \frac{\zeta_{\varphi\varphi}}{\sin^2\theta_m}\right).\tag{7}$$

Here, $\gamma = |\gamma_G|/(1+\alpha^2)$, in which γ_G is the gyromagnetic ratio and α the damping parameter; and $\zeta_{\theta\theta}$, $\zeta_{\varphi\varphi}$, $\zeta_{\varphi\theta}$, and $\zeta_{\theta\varphi}$ are the second derivatives of the magnetic free energy density, defined by the magnetization vector oriented by the angles θ_m and φ_m, at a given magnetic field.

Then, we performed fittings of the angular dependence of the resonance field H_r using Equation (6) and taking into account the free energy density for a system with uniaxial magnetic anisotropy combined with cubic magnetic anisotropy, i.e., Equation (5). The magnetic parameters m_s, k_u and k_{c1} obtained from the fits of the FMR data are summarized in Table 1.

Table 1. Parameters obtained from the fits of the experimental FMR data for our set of Co_2FeAl/W bilayers. For the fits, we assumed $\alpha \approx 2 \times 10^{-3}$ and $\gamma = 2.8\,MH/Oe$.

CFA Grown Temperature (K)	m_s (emu/cm^3)	k_u (ergs/cm^3)	k_{c1} (ergs/cm^3)
300	1340	1.00×10^5	3.01×10^5
573	1330	1.33×10^5	3.72×10^5
673	1330	1.66×10^5	4.82×10^5
773	1328	1.97×10^5	5.53×10^4

3.5. Experimental Results for Thermoelectric Effect

Based on the experimental results obtained in the structural and magnetic characterizations so far, from now, we focus our attention on the thermoelectric voltage measurements performed in the Co_2FeAl/W bilayers in which the CFA layer was deposited at the temperatures of 573 K and 673 K.

Figure 5 shows the experimental results of the magnetic response and the thermoelectric voltage for both samples. Specifically, Figure 5a,b presents the thermoelectric voltage as a function of the magnetic field, at selected φ values and setting $\Delta T = 27$ K. Notice the quite-interesting evolution in the shape of the curves as the magnitude and orientation of the field are altered. The one measured at $\varphi = 0°$ leads to a field configuration in which the thermoelectric voltage has a shape similar to that presented in the magnetization characterization, as we can clearly confirm from the plots shown in Figure 5c,d. However, with the increase of the φ, we verify a decrease of the thermoelectric voltage at saturation, leading to the suppression of the signal at $\varphi = 90°$, as expected. It is worth remarking that the overall thermoelectric voltage response is a result of the combination of effects associated with magnetic anisotropies in the samples and the thermoelectric voltage measurement configuration employed in the experiment [39].

Figure 5e,f shows V behavior as a function of φ, at $H = +500$ Oe, for selected ΔT values. At this field value, our system is magnetically saturated, in a sense that the magnetization follows the orientation of the magnetic field. As expected, the curves draw a clear dependence of V with φ, shown by a well-defined cosine shape. Furthermore, although the angular dependence is kept constant, the amplitude of the curves is altered with ΔT. All these features are in concordance with our theoretical predictions, discussed in the next sections.

3.6. Theoretical Approach for Thermoelectric Effect

As aforementioned, for ANE and LSSE effects, the application of a temperature gradient ∇T and a magnetic field $\vec{H}$ generates an electric field $\vec{E}$ given by Equation (1). Based on our experimental findings, for our theoretical approach, we considered a typical experiment in a film, in which the temperature gradient is normal to the film plane, while the magnetization lies in the plane, as depicted in Figure 1a.

The corresponding thermoelectric voltage, detected by electrical contacts at the ends of the main axis of the film, is thus given by

$$V = - \int_0^L \vec{E} \cdot d\vec{l},$$

(8)

where L is the distance between the electrical contacts and $d\vec{l} = dl\hat{j}$ in our case.

Then, using Equations (1) and (8), we obtain

$$V = \lambda \mu_o L \Delta T_f m_s cos\varphi_m.$$

(9)

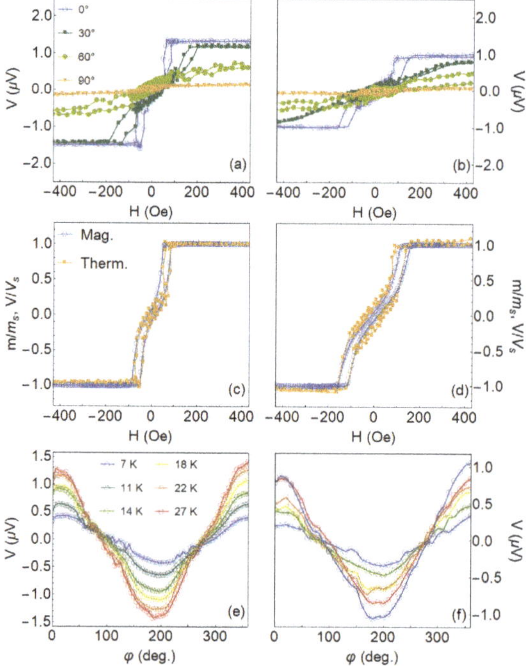

Figure 5. Experimental thermoelectric effect results. Thermoelectric voltage as a function of the magnetic field, at selected φ values and setting $\Delta T = 27$ K, for the (a) CFA/W bilayers in which the CFA layer was grown at (**a**) 573 K and (**b**) 673 K. (**c,d**) Comparison of the shape of normalized magnetization and thermoelectric voltage curves acquired at $\varphi = 0°$, with $\Delta T = 27$ K, for these samples. (**e,f**) Angular dependence of the thermoelectric voltage at $H = +500$ Oe for distinct ΔT values for the very same samples.

In this case, ΔT_f is the temperature variation across the bilayer, which is related with temperature variation ΔT measured experimentally across the sample [19,39],

$$\Delta T_f = \frac{t_f K_{sub}}{t_{sub} K_f} \Delta T,$$

(10)

where K_{sub} and t_{sub} are the thermal conductivity and thickness of the substrate, while K_f and t_f are the respective quantities for the CFA alloy. Here, we assumed $K_{sub} = 55$ W/Km and $t_{sub} = 0.7$ mm, while $K_f = 129$ W/Km and $t_f = 55$ nm for the CFA layer.

Notice that the V response given by Equation (9) is straightly dependent on the magnetic state of the sample through φ_m, which in turn is a result of the minimization process of free energy density of the system. This latter brings information on the magnetic field and the magnetic anisotropies in the system, given by Equation (5) in our case.

3.7. Numerical Calculations and Comparison between Theory and Experiment

We performed numerical calculations for two magnetic systems, distinct just with respect to the magnetic anisotropy dispersion. Remember that our system has a combination of induced uniaxial magnetic anisotropy and cubic magnetic anisotropy. Then, first, we calculated the magnetic and thermoelectric properties for a system without magnetic anisotropy dispersion, with $m_{si} = m_{s1} = ... =$

$m_{s50} = m_s = 1330$ G, $k_{ui} = k_{u1} = \ldots = k_{u50} = k_u$, with $h_k = 220$ Oe and $\varphi_{ki} = \varphi_{k1} = \ldots = \varphi_{k50} = 0°$ (x direction in Figure 1a), as well as $k_{c1i} = k_{c11} = \ldots = k_{c150} = 2.8\,k_u$. In a second moment, to mimic a magnetic system with a given uniaxial anisotropy dispersion, we considered the very same parameters, but with the fundamental difference of having φ_{ki} dispersed linearly around 0°, with dispersion range of $\Delta\varphi_k = 10°$.

Figure 6 shows the numerical calculations of the magnetic response and the thermoelectric voltage for systems having induced uniaxial magnetic anisotropy and cubic magnetic anisotropy, without and with uniaxial anisotropy dispersion.

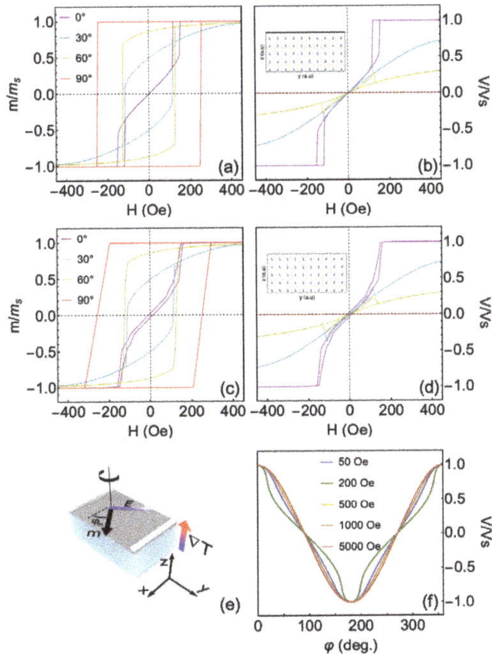

Figure 6. Numerical calculations of magnetic properties and thermoelectric voltage. (**a**) Normalized magnetization and (**b**) thermoelectric voltage curves, at selected φ values, for a system having induced uniaxial magnetic anisotropy and cubic magnetic anisotropy, without anisotropy dispersion. The calculations were performed with $m_{si} = m_{s1} = \ldots = m_{s50} = m_s = 1330$ G, $k_{ui} = k_{u1} = \ldots = k_{u50} = k_u$, with $h_k = 220$ Oe and $\varphi_{ki} = \varphi_{k1} = \ldots = \varphi_{k50} = 0°$, as well as $k_{c1i} = k_{c11} = \ldots = k_{c150} = 2.8\,k_u$. Furthermore, we considered $\Delta T = 27$ K. In (**b**), the inset illustrates a simple view of magnetic domains in the system without anisotropy dispersion. (**c,d**) Similar plots for the the same system having uniaxial anisotropy dispersion, in which φ_{ki} dispersed linearly around 0°, with dispersion range of $\Delta\varphi_k = 10°$. In (**d**), the inset also brings the simple view of the magnetic domains in the system without anisotropy dispersion. (**e**) Schematic representation of our bilayer structure in the experiment, in which we depict φ_m that is inserted in Equation (9) for the calculations. (**f**) Angular dependence of the thermoelectric voltage at $\Delta T = 27$ K for distinct field values for the system with anisotropy dispersion. The curves are normalized in order to make easier the direct comparison between results.

For the system without anisotropy dispersion (Figure 6a,b), we verify that the magnetization curve calculated for $\varphi = 0°$ reflects the behavior of an intermediate magnetic axis, a result of the competition between the uniaxial and cubic magnetic anisotropies. Furthermore, as expected, the easy and hard magnetization axes are aligned to $\varphi = 90°$ and 45°, respectively [46]. This behavior is in concordance with the experimental results previously discussed in Sections 3.2 and 3.4. Regarding the thermoelectric calculations, the results obtained for the $\varphi = 0°$ leads to a curve with a similar shape to that one obtained in the magnetization calculation. With the increase of φ, we confirm the decrease in the thermoelectric voltage at saturation, reaching zero for $\varphi = 90°$. These features are in perfect agreement with our discussion presented in Section 3.5.

When the magnetic anisotropy dispersion is inserted in the system (Figure 6c,d), the overall effective magnetic anisotropy is not strongly altered. As a consequence, the shapes of the curves in the magnetization response and thermoelectric voltage are globally kept. However, remarkably, the anisotropy dispersion refines the model, introducing details of the magnetic behavior that are fundamental to the description of real systems. Specifically, the inclusion of the anisotropy dispersion leads to smoother magnetization variations and is responsible for modifications in the magnetic permeability, coercive field and remanent magnetization. All these changes yield deep impacts in the thermoelectric voltage response.

As aforementioned, the thermoelectric voltage response is a combination of effects associated with magnetic anisotropies in the samples and the thermoelectric voltage measurement configuration employed in the experiment. In some sense, it is reasonable its dependence with φ (Figure 6e). Therefore, we also perform numerical calculations of the V behavior as a function of φ for the system with anisotropy dispersion, at distinct H values, shown in Figure 6f. As expected, the curves of the thermoelectric voltage are strongly dependent on both the φ angle and the amplitude of the magnetic field H. The equilibrium angle of magnetization, φ_m obtained from the minimization of Equation (5), is the result of the competition between the energy density contributions of magnetic induced uniaxial and cubic anisotropies with the Zeeman energy density. For field values high enough to saturate the system magnetically, the angular dependence of V is in perfect concordance with Equation (9), given by a cosine of φ, given $\varphi_m = \varphi$. However, for unsaturated states, the cosine behavior is lost, bringing information on the magnetic anisotropy and anisotropy field.

It is interesting to notice that the previous calculations have qualitatively described the main features of the magnetic behavior and thermoelectric voltage response in a system having induced uniaxial magnetic anisotropy combined with cubic magnetic anisotropy.

However, the most striking findings here are shown in Figure 7. Here we directly compare theory and experiment and uncover the evolution of the magnetic response and thermoelectric voltage with the amplitude and direction of the magnetic field. Furthermore, we corroborate the major role of the magnetic anisotropy dispersion to the description of the magnetic behavior and thermoelectric effect.

Notice the remarkable agreement between experiment and theory for the different conditions of measurement. Specifically, we disclose experimental results for the CFA/W bilayer in which the CFA layer was grown at 573 K. For the numerical calculations, we considered a system with uniaxial anisotropy dispersion, with $m_{si} = m_{s1} = \ldots = m_{s50} = m_s = 1330$ G, $k_{ui} = k_{u1} = \ldots = k_{u50} = k_u$, where $h_k = 220$ Oe and φ_{ki} is dispersed linearly around $0°$, with dispersion range of $\Delta\varphi_k = 10°$, as well as $k_{cli} = k_{cl1} = \ldots = k_{cl50} = 2.8\,k_u$. Remember that these magnetic parameters are the very same used for the previous calculations shown in Figure 6, all of them obtained from the magnetic characterization discussed in Sections 3.2 and 3.4. Therefore, from the comparison, we confirm the validity of our theoretical approach, including the description of our system given by Equation (5), as well as we corroborate magnetic parameters as the anisotropy constants and the saturation magnetization, presented in Table 1.

Hence, we were able to describe through the numerical calculations all the main features of the magnetic behavior and thermoelectrical effect in a magnetic system having induced uniaxial magnetic anisotropy combined with cubic magnetic anisotropy. From a general point of view, we highlight that this theoretical approach may be modified to describe any magnetic system if considered the appropriate free energy density. Furthermore, here, we showed the possibility of modulating the LSSE thermoelectric effect by the magnetic anisotropy induction. Specifically, the thermoelectric

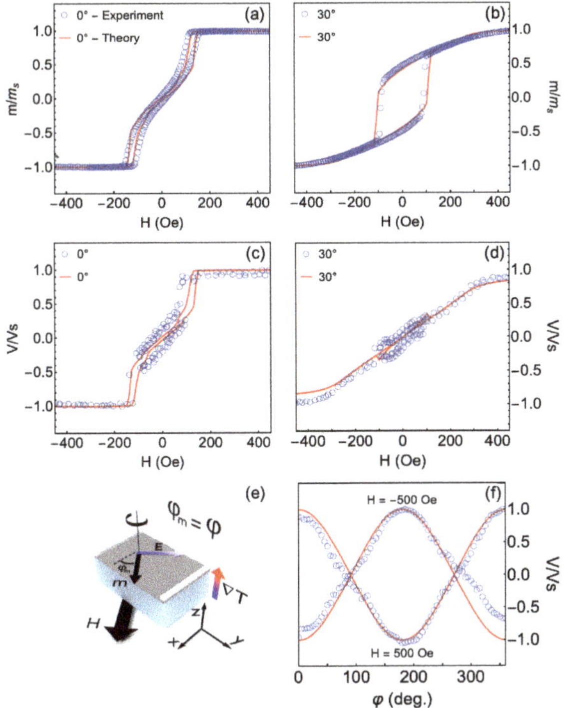

Figure 7. Comparison between theory and experiment: Normalized magnetization curves, at (**a**) $\varphi = 0°$ and (**b**) 30°, for the CFA/W bilayer in which the CFA layer was grown at 573 K. (**c**,**d**) Corresponding normalized thermoelectric voltage curves. (**e**) Schematic representation of our bilayer structure in the experiment, in which we depict φ_m that is inserted in Equation (9) for the calculations. (**f**) Normalized curves of the angular dependence of the thermoelectric voltage, at $\Delta T = 27$ K, for $H = \pm 500$ Oe. In particular, the calculations here were performed system with uniaxial anisotropy dispersion, with $m_{si} = m_{s1} = ... = m_{s50} = m_s = 1330$ G, $k_{ui} = k_{u1} = ... = k_{u50} = k_u$, where $h_k = 220$ Oe and φ_{ki} is dispersed linearly around 0°, with dispersion range of $\Delta\varphi_k = 10°$, as well as $k_{c1i} = k_{c11} = ... = k_{c150} = 2.8\,k_u$.

effect, measured through LSSE, emerged as a powerful tool to investigate fundamental parameters of a magnetic system, such as anisotropy configuration and anisotropy dispersion.

4. Conclusions

In conclusion, we investigated the Longitudinal Spin Seebeck Effect in films with a CFA/W bilayer structure grown onto GaAs (100) substrate. We verified that the magnetic properties of the CFA films are strongly dependent of parameters used during the production process, such as temperature deposition and annealing after the deposition. By setting optimal conditions, we manufactured systems having induced uniaxial magnetic anisotropy combined with cubic magnetic anisotropy. We observed clear dependence of the thermoelectric voltage curves with the magnetic anisotropy and experimental parameters employed in the LSSE experiment. By comparing experiment and theory, we confirmed the possibility of modulating a thermoelectric effect by magnetic anisotropy in Co_2FeAl Heusler alloy. Our results raise important features, contributing to the integration between electrical and magnetic properties that may promote improvements of the thermoelectric response in such nanostructures. These results enabled us to modulate the thermoelectric response as a function of the external magnetic field, an essential feature for sensor applications.

Author Contributions: M.V.L. and M.A.C. designed the experiments and conceived this study. M.V.L., E.C.d.S. and J.G.S. fabricated the samples. M.V.L. and E.C.d.S. performed the experimental measurements under supervision of A.B.d.O. and M.A.C., M.A.C., J.M.d.A., F.B. and L.L. performed the theoretical approach. The manuscript was written by M.A.C., M.V.L. and F.B. All authors have given approval for the final version of the manuscript.

Funding: This research was funded by CNPq: 407385/2018-5.

Acknowledgments: The authors acknowledge financial support from the Brazilian agencies CNPq and CAPES.

Conflicts of Interest: The authors declare no conflict of interest.

References

1. Raijmakers, L.; Danilov, D.; Eichel, R.A.; Notten, P. A review on various temperature-indication methods for Li-ion batteries. *Appl. Energy* **2019**, *240*, 918–945. [CrossRef]
2. Van Herwaarden, A.; Sarro, P. Thermal sensors based on the seebeck effect. *Sens. Actuators* **1986**, *10*, 321–346. [CrossRef]
3. Bali, A.; Chetty, R.; Mallik, R.C. Thin Film Thermoelectric Materials for Sensor Applications: An Overview. In *Thin Film Structures in Energy Applications*; Springer International Publishing: Cham, Switzerland, 2015.
4. Kirihara, A.; Kondo, K.; Ishida, M.; Ihara, K.; Iwasaki, Y.; Someya, H.; Matsuba, A.; Uchida, K.I.; Saitoh, E.; Yamamoto, N.; Kohmoto, S.; Murakami, T. Flexible heat-flow sensing sheets based on the longitudinal spin Seebeck effect using one-dimensional spin-current conducting films. *Sci. Rep.* **2016**, *6*, 23114. [CrossRef] [PubMed]
5. Uchida, K.; Ishida, M.; Kikkawa, T.; Kirihara, A.; Murakami, T.; Saitoh, E. Longitudinal spin Seebeck effect: From fundamentals to applications. *J. Phys. Condens. Matter* **2014**, *26*, 343202. [CrossRef] [PubMed]
6. Duan, Z.; Miao, B.; Sun, L.; Wu, D.; Du, J.; Ding, H. The Longitudinal Spin Seebeck Coefficient of Fe. *IEEE Magn. Lett.* **2019**, *10*, 1–5. [CrossRef]
7. Hou, D.; Qiu, Z.; Saitoh, E. Spin transport in antiferromagnetic insulators: progress and challenges. *NPG Asia Mater.* **2019**, *11*, 35. [CrossRef]
8. Sengupta, P.; Wen, Y.; Shi, J. Spin-dependent magneto-thermopower of narrow-gap lead chalcogenide quantum wells. *Sci. Rep.* **2018**, *8*, 5972. [CrossRef]
9. Mani, A.; Pal, S.; Benjamin, C. Designing a highly efficient graphene quantum spin heat engine. *Sci. Rep.* **2019**, *9*, 6018. [CrossRef]
10. Ando, K.; Morikawa, M.; Trypiniotis, T.; Fujikawa, Y.; Barnes, C.H.W.; Saitoh, E. Direct conversion of light-polarization information into electric voltage using photoinduced inverse spin-Hall effect in Pt/GaAs hybrid structure: Spin photodetector. *J. Appl. Phys.* **2010**, *107*, 113902. [CrossRef]
11. Isella, G.; Bottegoni, F.; Ferrari, A.; Finazzi, M.; Ciccacci, F. Photon energy dependence of photo-induced inverse spin-Hall effect in Pt/GaAs and Pt/Ge. *Appl. Phys. Lett.* **2015**, *106*, 232402. [CrossRef]
12. Zhang, Z.Q.; Yang, Y.R.; Fu, H.H.; Wu, R. Design of spin-Seebeck diode with spin semiconductors. *Nanotechnology* **2016**, *27*, 505201. [CrossRef]
13. Liao, T.; Ye, Z.; Chen, J. Spin-Seebeck Temperature Sensors. *IEEE Trans. Electron Devices* **2017**, *64*, 2655–2658. [CrossRef]
14. Bauer, G.E.W.; Saitoh, E.; van Wees, B.J. Spin caloritronics. *Nat. Mater.* **2012**, *11*, 391–399. [CrossRef]
15. Sakuraba, Y.; Hasegawa, K.; Mizuguchi, M.; Kubota, T.; Mizukami, S.; Miyazaki, T.; Takanashi, K. Anomalous Nernst Effect in $L1_0$-FePt/MnGa Thermopiles for New Thermoelectric Applications. *Appl. Phys. Express* **2013**, *6*, 033003. [CrossRef]
16. Zhang, B.; Meng, K.K.; Yang, M.Y.; Edmonds, K.; Zhang, H.; Cai, K.M.; Sheng, Y.; Zhang, N.; Ji, Y.; Zhao, J.H.; et al. Piezo voltage controlled planar Hall effect devices. *Sci. Rep.* **2016**, *6*, 28458. [CrossRef] [PubMed]

17.   Baxter, J.; Lesina, A.C.; Guay, J.M.; Weck, A.; Berini, P.; Ramunno, L. Plasmonic colours predicted by deep learning. *Sci. Rep.* **2019**, *9*, 8074. [CrossRef]

18.   Gueye, M.; Zighem, F.; Belmeguenai, M.; Gabor, M.; Tiusan, C.; Faurie, D. Effective 90-degree magnetization rotation in Co_2FeAl thin film/piezoelectric system probed by microstripline ferromagnetic resonance. *Appl. Phys. Lett.* **2015**, *107*, 032908. [CrossRef] [PubMed]

19.   Kikkawa, T.; Uchida, K.; Daimon, S.; Shiomi, Y.; Adachi, H.; Qiu, Z.; Hou, D.; Jin, X.F.; Maekawa, S.; Saitoh, E. Separation of longitudinal spin Seebeck effect from anomalous Nernst effect: Determination of origin of transverse thermoelectric voltage in metal/insulator junctions. *Phys. Rev. B* **2013**, *88*, 214403. [CrossRef]

20.   Holanda, J.; Alves Santos, O.; Cunha, R.O.; Mendes, J.B.; Rodríguez-Suárez, R.L.; Azevedo, A.; Rezende, S.M. Longitudinal spin Seebeck effect in permalloy separated from the anomalous Nernst effect: Theory and experiment. *Phys. Rev. B* **2017**, *95*, 214421. [CrossRef]

21.   Manipatruni, S.; Nikonov, D.E.; Lin, C.C.; Gosavi, T.A.; Liu, H.; Prasad, B.; Huang, Y.L.; Bonturim, E.; Ramesh, R.; Young, I.A. Scalable energy-efficient magnetoelectric spin–orbit logic. *Nature* **2019**, *565*, 35. [CrossRef]

22.   Trudel, S.; Gaier, O.; Hamrle, J.; Hillebrands, B. Magnetic anisotropy, exchange and damping in cobalt-based full-Heusler compounds: An experimental review. *J. Phys. D Appl. Phys.* **2010**, *43*, 193001. [CrossRef]

23.   Husain, S.; Kumar, A.; Kumar, P.; Kumar, A.; Barwal, V.; Behera, N.; Choudhary, S.; Svedlindh, P.; Chaudhary, S. Spin pumping in the Heusler alloy Co_2FeAl/MoS_2 heterostructure: Ferromagnetic resonance experiment and theory. *Phys. Rev. B* **2018**, *98*, 180404. [CrossRef]

24.   Akansel, S.; Kumar, A.; Behera, N.; Husain, S.; Brucas, R.; Chaudhary, S.; Svedlindh, P. Thickness-dependent enhancement of damping in Co_2FeAl/β-Ta thin films. *Phys. Rev. B* **2018**, *97*, 134421. [CrossRef]

25.   Boehnke, A.; Martens, U.; Sterwerf, C.; Niesen, A.; Huebner, T.; von der Ehe, M.; Meinert, M.; Kuschel, T.; Thomas, A.; Heiliger, C.; Münzenberg, M.; Reiss, G. Large magneto-Seebeck effect in magnetic tunnel junctions with half-metallic Heusler electrodes. *Nat. Commun.* **2017**, *8*, 1626. [CrossRef]

26.   Hu, S.; Itoh, H.; Kimura, T. Efficient thermal spin injection using CoFeAl nanowire. *NPG Asia Mater.* **2014**, *6*, e127-1–e127-5. [CrossRef]

27.   Inomata, K.; Ikeda, N.; Tezuka, N.; Goto, R.; Sugimoto, S.; Wojcik, M.; Jedryka, E. Highly spin-polarized materials and devices for spintronics. *Sci. Tecnhol. Adv. Mater.* **2008**, *9*, 014101. [CrossRef]

28.   Gabor, M.S.; Petrisor, Jr., T.; Tiusan, C.; Hehn, M.; Petrisor, T. Magnetic and structural anisotropies of Co_2FeAl Heusler alloy epitaxial thin films. *Phys. Rev. B* **2011**, *84*, 134413. [CrossRef]

29.   Lai, B.; Zhang, X.; Lu, X.; Yang, L.; Wang, J.; Chen, Y.; Zhao, Y.; Li, Y.; Ruan, X.; Wang, X.; et al. Magnetic anisotropy of half-metallic Co_2FeAl ultra-thin films epitaxially grown on GaAs (001). *AIP Adv.* **2019**, *9*, 065002. [CrossRef] [PubMed]

30.   Qiao, S.; Nie, S.; Zhao, J.; Zhang, X. Temperature dependent magnetic anisotropy of epitaxial Co_2FeAl films grown on GaAs. *J. Appl. Phys.* **2015**, *117*, 093904. [CrossRef]

31.   Xu, X.; Zhang, D.; Wang, W.; Wu, Y.; Wang, Y.; Jiang, Y. Surface effects on the magnetic properties of $Co_2FeAl(001)$: An ab initio study. *J. Magn. Magn. Mater.* **2010**, *322*, 3351–3354. [CrossRef]

32.   Yu, L.; Gao, G.; Zhu, L.; Deng, L.; Yang, Z.; Yao, K. First principles study of magnetoelectric coupling in $Co_2FeAl/BaTiO_3$ tunnel junctions. *Phys. Chem. Chem. Phys.* **2015**, *17*, 14986–14993. [CrossRef]

33.   Li, J.; Zhang, G.; Peng, C.; Wang, W.; Yang, J.; Wang, Y.; Cheng, Z. Magneto-Seebeck effect in $Co_2FeAl/MgO/Co_2FeAl$: first-principles calculations. *Phys. Chem. Chem. Phys.* **2019**, *21*, 5803–5812. [CrossRef]

34.   Xu, X.; Zhang, D.; Wu, Y.; Zhang, X.; Li, X.; Yang, H.; Jiang, Y. Electronic structures of Heusler alloy $Co_2FeAl_{1-x}Si_x$ surface. *Rare Met.* **2012**, *31*, 107–111. [CrossRef]

35.   Belmeguenai, M.; Tuzcuoglu, H.; Gabor, M.S.; Petrisor, T.; Tiusan, C.; Zighem, F.; Chérif, S.M.; Moch, P. Co_2FeAl Heusler thin films grown on Si and MgO substrates: Annealing temperature effect. *J. Appl. Phys.* **2014**, *115*, 043918. [CrossRef]

36.   Gueye, M.; Wague, B.M.; Zighem, F.; Belmeguenai, M.; Gabor, M.S.; Petrisor, Jr., T.; Tiusan, C.; Mercone, S.; Faurie, D. Bending strain-tunable magnetic anisotropy in Co_2FeAl Heusler thin film on Kapton®. *Appl. Phys. Lett.* **2014**, *105*, 062409-1–062409-4. [CrossRef]

37. Belmeguenai, B.; Tuzcuoglu, H.; Gabor, M.S.; Petrisor, Jr., T.; Tiusan, C.; Berling, D.; Zighem, F.; Chauveau, T.; Chérif, S.M.; Moch, P. Co$_2$FeAl thin films grown on MgO substrates: Correlation between static, dynamics, and structural properties. *Phys. Rev. B* **2013**, *87*, 184431. [CrossRef]

38. Silva, A.; Escobar, V.; Callegari, G.; Agra, K.; Chesman, C.; Bohn, F.; Corrêa, M. Giant magnetoimpedance effect in Co$_2$FeAl single layered and Co$_2$FeAl/Ag multilayered films in amorphous substrates. *Mat. Lett.* **2015**, *156*, 90–93. [CrossRef]

39. Melo, A.S.; de Oliveira, A.B.; Chesman, C.; Della Pace, R.D.; Bohn, F.; Correa, M.A. Anomalous nernst effect in stressed magnetostrictive film grown onto flexible substrate. *Sci. Rep.* **2019**, *9*, 15338. [CrossRef]

40. Silva, E.F.; Correa, M.A.; Della Pace, R.D.; Plá Cid, C.C.; Kern, P.R.; Carara, M.; Chesman, C.; Alves Santos, O.; Rodríguez-Suárez, R.L.; Azevedo, A.; Rezende, S.M.; Bohn, F.; et al. Thickness dependence of the magnetic anisotropy and dynamic magnetic response of ferromagnetic NiFe films. *J. Phys. D Appl. Phys.* **2017**, *50*, 185001. [CrossRef]

41. Janik, E.; Dłużewski, P.; Kret, S.; Presz, A.; Kirmse, H.; Neumann, W.; Zaleszczyk, W.; Baczewski, L.T.; Petroutchik, A.; Dynowska, E.; Sadowski, J.; Caliebe, W.; Karczewski, G.; Wojtowicz, T. Catalytic growth of ZnTe nanowires by molecular beam epitaxy: Structural studies. *Nanotechnology* **2007**, *18*, 475606. [CrossRef] [PubMed]

42. Cao, X.A.; Hu, H.T.; Dong, Y.; Ding, X.M.; Hou, X.Y. The structural, chemical, and electronic properties of a stable GaS/GaAs interface. *J. Appl. Phys.* **1999**, *86*, 6940–6944. [CrossRef] [PubMed]

43. Leo, G. Influence of a ZnTe buffer layer on the structural quality of CdTe epilayers grown on (100) GaAs by metalorganic vapor phase epitaxy. *J. Vac. Sci. Technol. B Microelectron. Nanom. Struct.* **1996**, *14*, 1739. [CrossRef]

44. De Teresa, J.M.; Serrate, D.; Cordoba, R.; Yusuf, S.M. Correlation between the synthesis conditions and the compositional and magnetic properties of Co$_2$(Cr$_{1-x}$Fe$_x$)Al Heusler alloys. *J. Alloy. Compd.* **2008**, *450*, 31–38. [CrossRef]

45. Belmeguenai, M.; Gabor, M.S.; Zighem, F.; Roussigné, Y.; Faurie, D.; Tiusan, C. Annealing temperature and thickness dependencies of structural and magnetic properties of Co$_2$FeAl thin films. *Phys. Rev. B* **2016**, *94*, 104424. [CrossRef]

46. Bezerra, C.; Chesman, C.; Albuquerque, E.; Azevedo, A. Effects of the magneto-crystalline anisotropy on the magnetic properties of Fe/Cr/Fe (110) trilayer. *Eur. Phys. J. B* **2004**, *39*, 527–533. [CrossRef]

47. Stoner, E.; Wohlfarth, E. A mechanism of magnetic hysteresis in heterogeneous alloys. *Phil. Trans. Roy.Soc.* **1948**, *240*, 599–642. [CrossRef]

48. Tao, X.; Wang, H.; Miao, B.; Sun, L.; You, B.; Wu, D.; Zhang, W.; Oepen, H.; Zhao, J.; Ding, H. Unveiling the Mechanism for the Split Hysteresis Loop in Epitaxial Co$_2$Fe$_{1-x}$Mn$_x$Al Full-Heusler Alloy Films. *Sci. Rep.* **2016**, *6*, 18615. [CrossRef]

49. Pandey, H.; Joshi, P.; Pant, R.P.; Prasad, R.; Auluck, S.; Budhani, R. Evolution of ferromagnetic and spin-wave resonances with crystalline order in thin films of full-Heusler alloy Co$_2$MnSi. *J. Appl. Phys.* **2012**, *111*, 023912. [CrossRef]

Article

The Performance of the Magneto-Impedance Effect for the Detection of Superparamagnetic Particles

Alfredo García-Arribas [1,2]

[1] Departamento de Electricidad y Electrónica, Universidad del País Vasco UPV/EHU, 48940 Leioa, Spain;
    alfredo.garcia@ehu.es
[2] Basque Centre for Materials, Applications and Nanostructures, BCMaterials, 48940 Leioa, Spain

Received: 4 March 2020; Accepted: 29 March 2020; Published: 31 March 2020

Abstract: The performance of magneto-impedance sensors to detect the presence and concentration of magnetic nanoparticles is investigated, using finite element calculations to directly solve Maxwell's equations. In the case of superparamagnetic particles that are not sufficiently magnetized by an external field, it is assumed that the sensitivity of the magneto-impedance sensor to the presence of magnetic nanoparticles comes from the influence of their magnetic permeability on the sensor impedance, and not from the stray magnetic field that the particles produce. The results obtained not only justify this hypothesis, but also provide an explanation for the discrepancies found in the literature about the response of magneto-impedance sensors to the presence of magnetic nanoparticles, where some authors report an increasing magneto-impedance signal when the concentration of magnetic nanoparticles is increased, while others report a decreasing tendency. Additionally, it is demonstrated that sensors with lower magneto-impedance response display larger sensitivities to the presence of magnetic nanoparticles, indicating that the use of plain, nonmagnetic conductors as sensing materials can be beneficial, at least in the case of superparamagnetic particles insufficiently magnetized in an external magnetic field.

Keywords: magneto-impedance; biosensor; finite-element method

---

1. Introduction

Magnetic nanoparticles (MNPs) are used in numerous biomedical applications, both for diagnosis and therapy [1]. In diagnosis, they are used for magnetic resonance imaging enhancement [2] and, after adequate functionalization, for magnetic separation, concentration, and detection of specific analytes [3]. In therapy, MNPs are used in drug delivery and for hyperthermia treatments, among others [4]. The most relevant type of magnetic particles for these applications are superparamagnetic iron oxide nanoparticles, SPIONs. In order to detect the presence of MNPs and to quantify their concentration, different types of magnetic sensors have been proposed [5,6]. In particular, the magneto-impedance (MI) effect's extraordinary sensitivity to small magnetic fields has motivated the active development of MI-based biosensors which, fundamentally, detect the presence and concentration of magnetic particles [7–13].

The MI effect consists in the large variation of the electrical impedance Z of a soft magnetic conductor when subjected to an external magnetic field [14]. The phenomenon can be understood from the classical electromagnetic theory as a consequence of the skin effect's dependence on the permeability μ of the material. The skin effect—that is, the limited penetration of the alternating electromagnetic field in a conductor, is characterized by the penetration depth δ given by

$$\delta = (\pi f \sigma \mu)^{-1/2} \tag{1}$$

where f is the frequency of the field, and σ the conductivity of the material. Depending on the magnetic behavior of the sample, the external magnetic field modifies the permeability, which subsequently changes the effective electromagnetic cross section of the conductor, and produces concomitant variations in the impedance. Figure 1 shows a sketch of a typical MI curve in a planar sample with transverse anisotropy. The maximum impedance $Z_{\max}$ is obtained for an applied field similar in magnitude to the anisotropy field H^k, when the transverse permeability reaches its maximum value $\mu_{\max}$. The minimum impedance $Z_{\min}$ is obtained when the sample is magnetically saturated and the permeability $\mu_{\min}$ is close to μ_0.

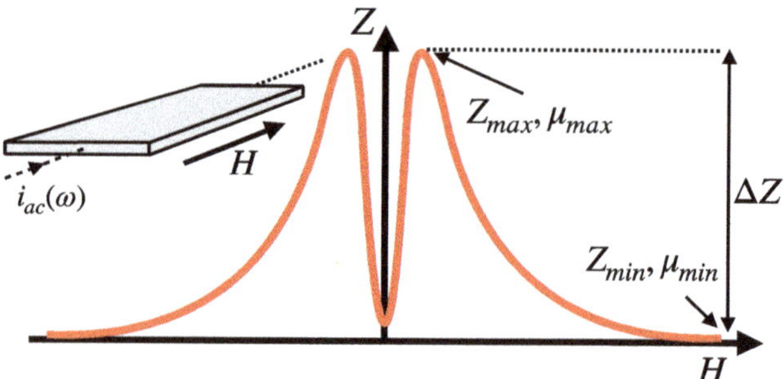

Figure 1. Sketch of the dependence of the impedance on the applied magnetic field in a soft magnetic sample with transverse anisotropy (in-plane easy axis, perpendicular to the current flow and the applied field).

In this kind of sample, the largest sensitivity to the external magnetic field is reached in the region between $H = 0$ and at the peak where $H = H^k$. For instance, the sensitivity of a multilayered planar sample can reach a value of 27 kΩ/T, occurring at an applied field of $\mu_0 H = 200$ μT ($H = 180$ A/m) and measured at 23 MHz [15].

The MI is usually quantified as the ratio of impedance change, and its maximum value (at each frequency), given by [14]

$$\text{MI}_{\max}(\%) = \frac{Z_{\max} - Z_{\min}}{Z_{\min}} \times 100 \tag{2}$$

is usually used as a figure of merit for MI sensors. In the literature, when used for detecting the presence and concentration of MNPs, it is usual to report the changes produced by the presence of nanoparticles in the complete MI curve and in particular, in its maximum value $\text{MI}_{\max}$. Wang and collaborators [16] have compiled the results from quite a number of works that use the MI effect to detect magnetic particles. Overall, the comparison of the results seems to be utterly inconsistent, because some authors report an increase in the MI response when increasing the concentration of particles (for example, Devkota and collaborators [17]), while others describe a decreasing tendency (for example, Yang and collaborators [18]). As the MI materials and the nature of the particles differ between studies, it becomes necessary to adopt a systematic approach to evaluate the sensitivity of the MI effect to quantify the concentration of MNPs. The present work attempts to shed light on this issue from a theoretical point of view, using a numerical procedure (finite element method) to solve Maxwell's equations and calculate the impedance in the conditions of the experiments described in the literature.

An important point is that the authors of the mentioned works compiled in [16], in which contradicting results are reported, usually do not provide a satisfactory explanation regarding the physical origin of the MI effect's sensitivity to the presence of magnetic nanoparticles. It is assumed that it is a consequence of the MI effect's extraordinary sensitivity to low magnetic fields. The fact, however, is that superparamagnetic particles do not display remanence, and therefore do not produce external magnetic fields unless magnetized by an appropriate biasing field. However, a large applied magnetic field saturates the MI sensor and drastically reduces its sensitivity. In these circumstances, the sensitivity of the MI sensor to the presence and concentration of the particles cannot be caused by the magnetic field that they produce. Of course, this is true only if the particle system shows true superparamagnetic behavior. Agglomeration, or a large particle-size distribution, can produce a finite remanence which affects the MI signal of the sensor. The analysis of experimental data becomes difficult without a detailed particle characterization. There are some works in which the MI effect has been properly used to quantify the concentration of superparamagnetic nanoparticles in a clever configuration by measuring their stray field when magnetized by a strong external field [10], but in most of the works that claim the detection of magnetic nanoparticles using the MI effect, there is no satisfactory explanation of the involved mechanism.

In this work, we make the more realistic assumption that it is the permeability of the magnetic nanoparticles that produces the change in the MI response of the sensor. The sensing mechanism is therefore very simple: the presence of a high permeability medium in the proximity of the MI sensor modifies the distribution of the electromagnetic field associated with the alternating current flowing in the sensor. This produces a variation of the sensor's impedance, but it is not related to the intrinsic sensitivity of the MI to low magnetic fields.

As explained in the next section, this concept is implemented in this work in the calculation of the MI response. The results from the numerical solution of Maxwell's equations not only confirm that our assumption qualitatively reproduces the experimental results, but also help to explain the discrepancies reported in the literature compiled in [16], about the MI behavior when detecting magnetic nanoparticles.

2. Numerical Calculation Procedure

We aim to evaluate the response of a magneto-impedance sensor in the presence of a sample containing superparamagnetic particles. The electrical impedance of a conductor can in principle, be analytically calculated using Maxwell equations. In practice, only in very simple cases with highly symmetric geometries can a closed expression be derived, usually with the use of severe approximations. For instance, in planar samples, infinite width and length is assumed to calculate the impedance, producing an expression as [19]:

$$Z = R_{dc}\sqrt{j\theta}\coth\sqrt{j\theta} \tag{3}$$

for a sample of thickness $2a$, where R_{dc} is the dc resistance, $j = \sqrt{-1}$, the imaginary unit, and $\theta = a\sqrt{2\pi f\sigma\mu} = \sqrt{2}a/\delta$. Similar expressions are obtained for the case of cylindrical samples (wires).

Using this type of expression, the magneto-impedance curve can be computed by incorporating the field dependence of the permeability, using models for the magnetization process and the dynamical behavior of magnetization [20,21]. Although useful, these models systematically overestimate the magnitude of the MI effect [22,23] when compared with actual measurements, mainly because the real conditions of the experimental setup are not usually considered. In particular, the contribution of the measuring circuit to the total impedance is not usually considered in the models.

Numerical simulation using Finite Element Methods (FEM) can overcome these limitations as the elements of the measuring circuit can be easily incorporated in the model. After specifying the material properties and the adequate boundary conditions, the impedance can be calculated by numerically solving Maxwell's equations for an arbitrary geometry. If the dependence of the material properties, especially the permeability, on the magnetic field H are known, the complete $Z(H)$ can be calculated. If only the performance (expressed as the MI ratio in Equation (2)) is of interest, FEM can be used to calculate the impedance for two values of permeability: μ_{max}, corresponding to the maximum value of transverse permeability (and the impedance Z_{max}) and $\mu_{min} = \mu_0$, corresponding to the saturated state (and the impedance Z_{min}) as depicted in Figure 1.

The versatility of FEM makes it possible to calculate the impedance of the MI sample in a variety of environments accurately resembling the experimental conditions. For the purpose of this work, using FEM, we calculate the impedance of an MI material in the presence of a sample of magnetic nanoparticles which is modeled as a homogeneous continuum material with a given permeability μ_P. As explained before, the permeability of the particles is the only relevant magnetic property in this approach. There is no attempt to model the individual magnetic behavior of the particles. To reproduce the conditions of the experiments found in the literature, the MI response is evaluated as a function of the concentration of the nanoparticles. If we assume that the nature of the particles is always the same, the mean permeability of the sample of magnetic nanoparticles is simply proportional to its concentration. That is, we can calculate the evolution of the MI sensor's response in the presence of a sample of magnetic nanoparticles with different concentrations by simply modifying the value of μ_P—the permeability of the MNPs system.

For the numerical calculations, the free 2D (two-dimensional) software package FEMM [24] is used. Its implementation is restricted to low-frequency electromagnetic problems, ignoring the displacement current. Therefore, it cannot simulate propagating effects or account for the dielectric properties of the materials. However, it can accurately resolve MI problems in which the important effects are the skin effect and the magneto-inductive effect [25].

As previously discussed, the contribution from the measuring circuit to the impedance must be considered to obtain realistic results. For that reason, the simulation is performed with the MI sample inserted in a microstrip transmission line, which is a popular text fixture for measuring MI in planar strips. Figure 2a schematizes the setup of the simulated experiment. The sample of magnetic nanoparticles is placed in the shape of a drop on the MI sensor in the microstrip transmission line. The sketch Figure 2b illustrates the layout of the 2D finite element problem. The simulation domain represents the middle plane of the real problem. Due to the symmetry, only half of it needs to be simulated. The simulation domain (including the space surrounding the microstrip line) is kept small to reduce computation time. Built-in FEM open boundary conditions are imposed on the boundary to guarantee correctness of the solution.

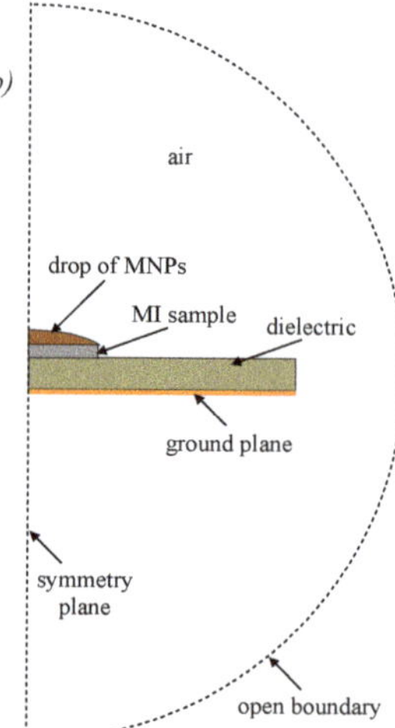

Figure 2. (a) Illustration of the experiment simulated by FEM: a drop containing a certain concentration of magnetic nanoparticles (MNPs) is placed on the MI material, which is inserted in a microstrip line to determine its impedance. (b) Layout (not at real-scale) of the two-dimensional problem solved by FEM, which corresponds to the middle plane of the setup as indicated in (a). Only half of the plane needs to be simulated due to symmetry.

The drop of magnetic nanoparticles is modeled as a nonconducting, homogeneous continuum material with a given permeability μ_P. As previously discussed, increasing values of μ_P correspond to increasing concentrations of MNPs.

The MI sample is modeled as a homogeneous conductor ($\sigma = 6.6 \times 10^5$ S/m) with a constant permeability μ. No definite magnetization process is assumed, and the MI is calculated as the relative ratio between the values of the impedance obtained with a high value of permeability μ_{max} and with a value of $\mu_{min} = \mu_0$. The geometry of the MI sample is kept constant through all the simulations: 1 mm wide and 20 μm thick (the length is not relevant for the 2D problem, and the net value of the impedance is calculated for a sample 1 m long).

In the microstrip line, the ground plane is modeled from pure cooper ($\sigma = 5.8 \times 10^7$ S/m, $\mu = \mu_0$) with a thickness of 35 μm. The 0.8 mm thick dielectric presents no conductivity, and $\mu = \mu_0$.

To calculate the impedance, an alternating current of a given frequency is imposed to flow through the sample, perpendicular to the plane of simulation. The current returns in the opposite direction through the ground conductor.

In this work, the MI is calculated for a large number of configurations with different values of μ_{max} and μ_P. For each configuration, the MI is calculated as a function of the frequency, in a range from 0 to 150 MHz or 0 to 1 GHz, depending on the case. To accommodate the intensive computational resources needed, we have made use of the XFEMM implementation of the software [26], which is run in a computer cluster.

3. Results and Discussion

Let us consider first the case of the MI sensor without any MNPs. Figure 3 presents the calculated MI response as a function of the frequency for a sample 1 mm wide and 20 μm thick. With the value of $\mu_{max} = 5000 \, \mu_0$ used in the simulation, the maximum value of MI ratio is $MI_{max} = 568\%$. The inset in Figure 3 shows the experimentally measured MI ratio of an amorphous ribbon composed of $Co_{65}Fe_4Ni_2Si_{15}B_{14}$. We can observe that the shapes of both curves are essentially similar, although the MI experimental values are significantly lower, indicating that the permeability $\mu_{max} = 5000 \, \mu_0$ used in the simulation is largely overestimated. Nevertheless, this case, which corresponds to a very sensitive MI sensor, is considered the starting point in our goal of studying the relevance of MI to detect magnetic nanoparticles.

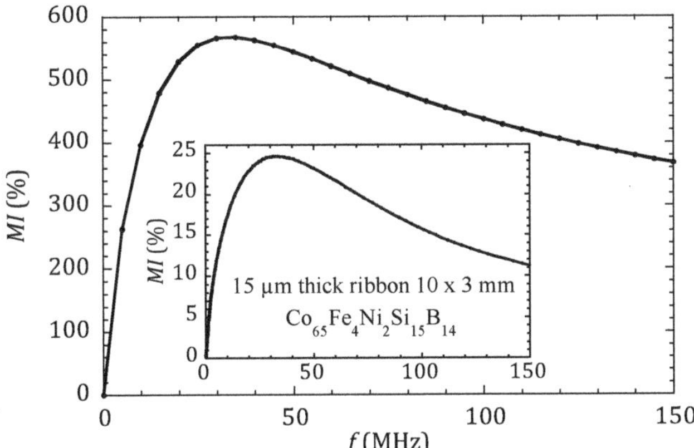

Figure 3. Magneto-impedance ratio MI as defined in Equation (2), calculated using FEM for the case of a planar sample 1 mm wide and 20 μm thick, with the permeability $\mu_{max} = 5000 \, \mu_0$. For comparison, the inset shows the MI ratio experimentally measured in an amorphous ribbon.

When a drop containing MNPs is placed on top of the MI sensor as described in Figure 2, the complete MI curve as a function of the frequency is modified. Figure 4 shows the variation of the region of the MI curves around the maximum when the concentration of MNPs is monotonically increased (increasing values of μ_P). The simulation results are divided into two plots for better clarity. Figure 4a determines that the MI response decreases when the permeability of the MNPs system increases from 2 μ_0 to 15 μ_0. However, when μ_P is increased further, Figure 4b shows that the MI response changes tendency and start to increase. In both plots, the large red and black dots indicate the position of MI_{max} for the different values of μ_P. Figure 5 compiles these results, plotting the evolution of MI_{max} for the whole range of tested μ_P values.

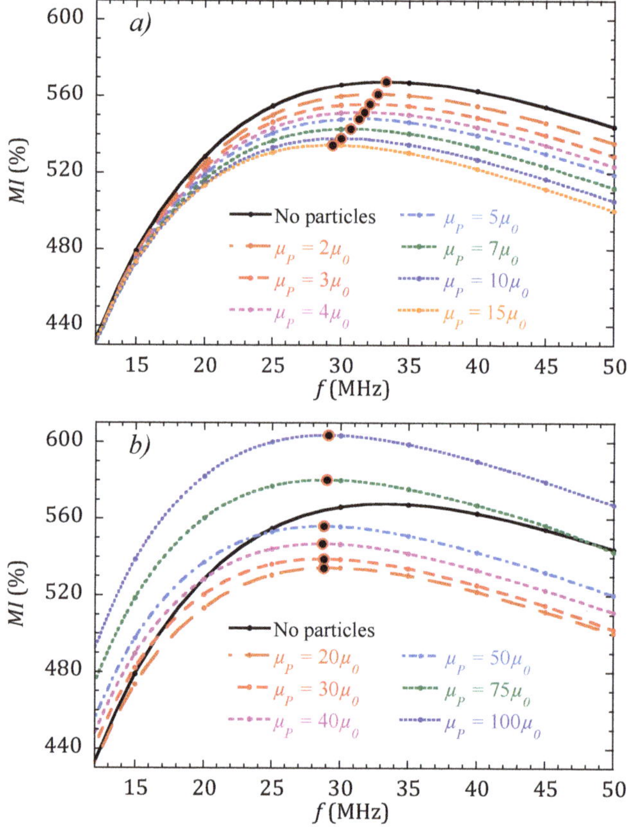

Figure 4. Variation of the MI curves calculated for a sensor with $\mu_{max} = 5000\ \mu_0$ when a drop with increasing MNPs concentration is placed on it. (**a**) for concentrations up to $\mu_P = 15\ \mu_0$, the magnitude of MI decreases. (**b**) For larger concentrations, from $\mu_P = 20\ \mu_0$, the magnitude of MI increases. Large red and black dots indicate the position of MI_{max} for each curve.

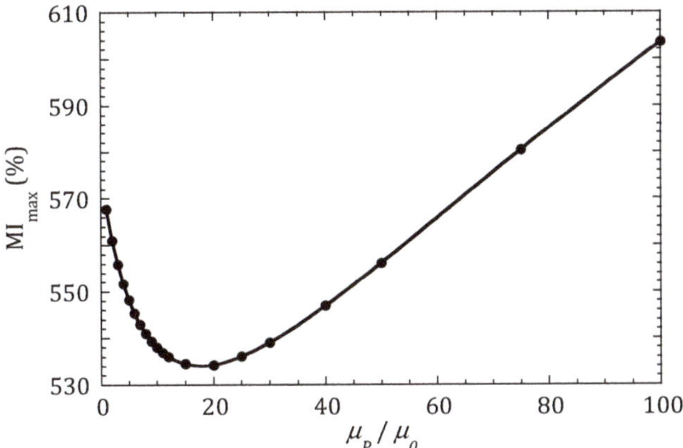

Figure 5. Maximum MI ratio of a sensor with $\mu_{\max} = 5000\ \mu_0$ when a drop of MNPs is placed on it, as a function of the concentration of nanoparticles (expressed as increasing permeability μ_P values).

Figure 5 shows that depending on the value of the permeability of the system of MNPs, the sensitivity of a very sensitive MI sensor can have a negative or positive value. That is, the MI ratio can display either a decreasing (for $\mu_P < 15$) or an increasing (for $\mu_P > 20$) behavior. This can certainly explain the discrepancies found in the compiled works [16] indicated in the introduction: the situation in the works where a decrease in the MI ratio is reported when increasing the concentration of MNPs, according to the results shown in Figure 5, can correspond to situations in which the permeability of the nanoparticle ensemble is low (due to a low intrinsic permeability of the nanoparticles or low concentrations). In contrast, the situation in the works reporting an increasing MI response with concentration can correspond to cases in which the permeability of the MNPs system is large.

As the maximum impedance ratio $\mathrm{MI}_{\max}$ is calculated as a quotient as displayed in Equation (2), the negative slope of the curve in Figure 5 for $\mu_P < 15$ indicates that $Z_{\min}$ increases more than $(Z_{\max} - Z_{\min})$ in this range. This is easily explained considering that the increase in the impedance is due to the presence of a magnetic medium (the particles) near the MI conductor. It is the same phenomenon occurring when a soft ferrite increases the impedance of a wire in a RF choke. When the permeability of the medium is very low compared with the permeability of the conductor itself, as in $Z_{\max}$, the influence of the magnetic medium is low. However, when the permeability of the conductor is low, as in $Z_{\min}$, even a surrounding medium with a low permeability produces a change in the impedance.

It is expected that the intrinsic MI performance of the sensor must have an influence on its sensitivity to the presence on MNPs. The same type of calculations that produced Figures 4 and 5 for the case of an MI sensor with $\mu_{\max} = 5000\ \mu_0$ have been performed for sensors with $\mu_{\max}/\mu_0 = 500$, 100, and 10—that is, with a decreasing intrinsic MI response (intrinsic $\mathrm{MI}_{\max}$ values are 156%, 46%, and 5.5%, respectively). We introduce a new parameter, η, to compare the capacity of the different sensors to detect MNPs with increasing concentrations, defined as

$$\eta(\%) = \frac{\mathrm{MI}_{\max}(\mu_P) - \mathrm{MI}_{\max,\mathrm{woP}}}{\mathrm{MI}_{\max,\mathrm{woP}}} \times 100, \tag{4}$$

where $\mathrm{MI}_{\max,\mathrm{woP}}$ is the maximum value of the MI ratio in the sensor without particles (the intrinsic $\mathrm{MI}_{\max}$ value). The parameter η therefore quantifies the sensitivity of an MI sensor to the presence of MNPs, expressed as the change in the MI ratio experienced by the sensor when a drop of MNPs is placed on top of it. Figure 6 plots the results obtained for η in the sensors with different MI performance (represented by their values of $\mu_{\max}$). Note that the curve for the sensor with $\mu_{\max} = 5000\ \mu_0$ is the same as the one plotted in Figure 5.

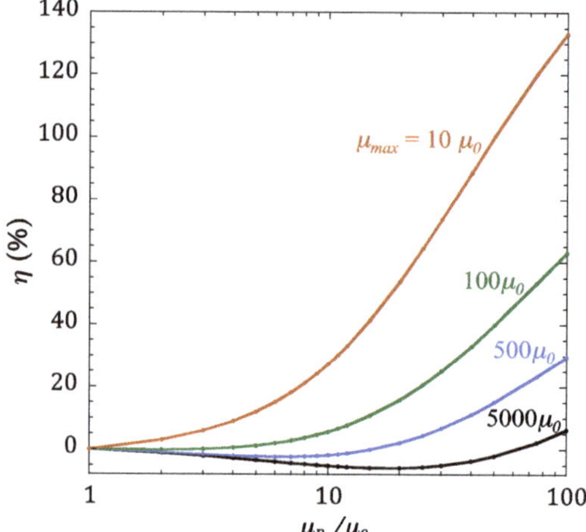

Figure 6. Relative change of the $MI_{\max}$ response, experienced by sensors with different MI performance (different values of $\mu_{\max}$) when a system of MNPs with increasing concentration is placed on top.

The data in Figure 6 is certainly conclusive: the sensors with worse MI performance are more sensitive to the presence of MNPs. If extrapolated, the best sensor material for detecting NMPs should be one with no MI effect at all. This is not completely surprising, as there is experimental evidence of a larger response from a plain Cu sensor than from an MI amorphous ribbon in a measurement performed in the same conditions [27]. In fact, this behavior could have been anticipated: if the change in the impedance of the sensor in the presence of MNPs is a consequence of changes in the distribution of the electromagnetic field due to the permeability of the particles, then this effect should be larger when the permeability of the sensor itself is smaller than that of the system of MNPs. The approach of using the impedance changes of nonmagnetic conductors to quantify the concentration of MNPs is being used successfully in the development of magnetic biosensors [28].

In accepting these conclusions, one should be aware of the limitations of the present analysis. The results from the FEM calculations do not include many relevant physical effects that occur in real systems, such as the dependence of the permeability on the frequency, or the possible agglomeration effects that takes place when the particle concentration is high, which completely change the magnetic behavior of the ensemble. In particular, the latter effect is difficult to model, but probably will modify the aspect of Figures 5 and 6 in which the MI sensitivity continuously increases with an increase in particle concentration.

Finally, it should be reminded that the analysis presented here was made for the case of superparamagnetic nanoparticles that do not produce stray magnetic fields unless magnetized. Highly sensitive magnetic sensors can still be used to detect larger monodomain nanoparticles presenting a net magnetic moment [29]. The modeling of the MI signal produced by ferromagnetically behaving particles requires a different approach to the one used here [30].

4. Conclusions

The performance of MI sensors to detect the presence and concentration of magnetic nanoparticles is investigated using finite element calculations to directly solve Maxwell's equations. The assumption is made that the sensitivity to MNPs comes from their magnetic permeability, and not from the stray magnetic field that they produce. This should be the case for superparamagnetic particles if they are not magnetized by an external magnetic field of sufficient strength. In this hypothesis, the variation of impedance experienced by the MI sensor is a consequence of the change in distribution of the electromagnetic field due to the permeability of the MNPs system.

The results obtained not only confirm that the sensitivity of MI sensors can be justified considering only the permeability of MNPs, but also help explain the origin of the discrepancies found in the literature about the response of magneto-impedance sensors to the presence of magnetic nanoparticles, where some authors report an increasing MI signal when the concentration of MNPs is increased, while others report a decreasing tendency. The results show that the change of the MI response when increasing the concentration of MNPs can be positive or negative, depending on the effective permeability of the particle system.

Additionally, the study demonstrated that a good intrinsic MI performance of the sensor has a detrimental effect on its capacity to detect MNPs: sensors with a lower MI response display larger sensitivities to the presence of magnetic nanoparticles. This seems to confirm that the correct strategy for detecting MNPs is to use the impedance change of plain, nonmagnetic conductors as sensing materials, at least in the case of superparamagnetic particles that are not magnetized in an external magnetic field.

Funding: This work is supported by the Spanish government (Project: MAT2017-83632-C3) and by the Basque government under grants KK-2019/00101 and IT1245-19.

Acknowledgments: SGI/IZO-SGIker (UPV/EHU) is acknowledged for allocation of computational resources and support.

References

1. Pankhurst, Q.A.; Connolly, J.; Jones, S.K.; Dobson, J. Applications of magnetic nanoparticles in biomedicine. *J. Phys. D Appl. Phys.* **2003**, *36*, R167–R181. [CrossRef]
2. Bao, Y.; Sherwood, J.A.; Sun, Z. Magnetic iron oxide nanoparticles as T-1 contrast agents for magnetic resonance imaging. *J. Mater. Chem. C* **2018**, *6*, 1280–1290. [CrossRef]
3. Angulo-Ibanez, A.; Eletxigerra, U.; Lasheras, X.; Campuzano, S.; Merino, S. Electrochemical tropomyosin allergen immunosensor for complex food matrix analysis. *Anal. Chim. Acta* **2019**, *1079*, 94–102. [CrossRef] [PubMed]
4. Saeed, M.; Ren, W.Z.; Wu, A.G. Therapeutic applications of iron oxide based nanoparticles in cancer: Basic concepts and recent advances. *Biomater. Sci.* **2018**, *6*, 708–725. [CrossRef]
5. Haun, J.B.; Yoon, T.J.; Lee, H.; Weissleder, R. Magnetic nanoparticle biosensors. *WIREs Nanomed. Nanobiotechnol.* **2010**, *2*, 291–304. [CrossRef]
6. Nabaei, V.; Chandrawati, R.; Heidari, H. Magnetic biosensors: Modeling and simulation. *Biosens. Bioelectron.* **2018**, *103*, 69–86. [CrossRef]
7. Kurlyandskaya, G.V.; Sánchez, M.L.; Hernando, B.; Prida, V.M.; Gorria, P.; Tejedor, M. Giant-magnetoimpedance-based sensitive element as a model for biosensors. *Appl. Phys. Lett.* **2003**, *82*, 2053–2055. [CrossRef]
8. García-Arribas, A.; Martínez, F.; Fernández, E.; Ozaeta, I.; Kurlyandskaya, G.V.; Svalov, A.V.; Berganzo, J.; Barandiaran, J.M. GMI detection of magnetic-particle concentration in continuous flow. *Sens. Actuators A Phys.* **2011**, *172*, 103–108. [CrossRef]
9. Devkota, J.; Howell, P.; Mukherjee, P.; Srikanth, H.; Mohapatra, S.; Phan, M.H. Magneto-reactance based detection of MnO nanoparticle-embedded Lewis lung carcinoma cells. *J. Appl. Phys.* **2015**, *117*, 17D123. [CrossRef]
10. Fodil, K.; Denoual, M.; Dolabdijan, C.; Treizebre, A.; Senez, V. In-flow detection of ultra-small magnetic particles by an integrated giant magnetic impedance sensor. *Appl. Phys. Lett.* **2016**, *108*, 173701. [CrossRef]
11. Wang, T.; Chen, Y.Y.; Wang, B.C.; He, Y.; Li, H.Y.; Liu, M.; Rao, J.J.; Wu, Z.Z.; Xie, S.R.; Luo, J. A giant magnetoimpedance-based separable-type method for supersensitive detection of 10 magnetic beads at high frequency. *Sens. Actuators A Phys.* **2019**, *300*, 111656. [CrossRef]

12. Beato, J.; Pérez-Landazábal, J.; Gómez-Polo, C. Enhanced magnetic nanoparticle detection sensitivity in non-linear magnetoimpedance-based sensor. *IEEE Sens. J.* **2018**, *18*, 8701–8708. [CrossRef]

13. Buznikov, N.A.; Safronov, A.P.; Orue, I.; Golubeva, E.V.; Lepalovskij, V.N.; Svalov, A.V.; Chlenova, A.A.; Kurlyandskaya, G.V. Modelling of magnetoimpedance response of thin film sensitive element in the presence of ferrogel: Next step toward development of biosensor for in-tissue embedded magnetic nanoparticles detection. *Biosens. Bioelectron.* **2018**, *117*, 366–472. [CrossRef]

14. Phan, M.H.; Peng, H.X. Giant magnetoimpedance materials: Fundamentals and applications. *Prog. Mater. Sci.* **2008**, *53*, 323–420. [CrossRef]

15. García-Arribas, A.; Fernández, E.; Svalov, A.; Kurlyandskaya, G.V.; Barandiaran, J.M. Thin-film magneto-impedance structures with very large sensitivity. *J. Magn. Magn. Mater.* **2016**, *400*, 321–326. [CrossRef]

16. Wang, T.; Zhou, Y.; Lei, C.; Luo, J.; Xie, S.; Pu, H. Magnetic impedance biosensor: A review. *Biosens. Bioelectron.* **2017**, *90*, 418–435. [CrossRef]

17. Yang, Z.; Liu, Y.; Lei, C.; Sun, X.C.; Zhou, Y. Ultrasensitive detection and quantification of *E. coli* O157:H7 using a giant magnetoimpedance sensor in an open-surface microfluidic cavity covered with an antibody-modified gold surface. *Microchim. Acta* **2016**, *183*, 1831–1837. [CrossRef]

18. Devkota, J.; Wang, C.; Ruiz, A.P.; Mohapatra, S.; Mukherjee, P.; Srikanth, H.; Phan, M.H. Detection of low-concentration superparamagnetic nanoparticles using an integrated radio frequency magnetic biosensor. *J. Appl. Phys.* **2013**, *113*, 104701. [CrossRef]

19. Chen, D.-X.; Muñoz, J.L. AC impedance and circular permeability of slab and cylinder. *IEEE Trans. Magn.* **1999**, *35*, 1906–1923. [CrossRef]

20. Makhnovskiy, D.P.; Panina, L.V.; Mapps, D.J. Field-dependent surface impedance tensor in amorphous wires with two types of magnetic anisotropy: Helical and circumferential. *Phys. Rev. B* **2001**, *63*, 144424. [CrossRef]

21. Kraus, L. Theory of giant magneto-impedance in the planar conductor with uniaxial magnetic anisotropy. *J. Magn. Magn. Mater.* **1999**, *195*, 764–778. [CrossRef]

22. Kraus, L. The theoretical limits of giant magneto-impedance. *J. Magn. Magn. Mater.* **1999**, *196*, 354–356. [CrossRef]

23. Barandiarán, J.M.; García-Arribas, A.; Muñoz, J.L.; Kurlyandskaya, G.V. Influence of magnetization processes and device geometry on the GMI effect. *IEEE Trans. Magn.* **2002**, *38*, 3051–3056. [CrossRef]

24. Meeker, D.C. Finite Element Method Magnetics, Version 4.2. Available online: http://www.femm.info (accessed on 28 February 2018).

25. García-Arribas, A.; Fernández, E.; De Cos, D. Thin-film magneto-impedance sensors. In *Magnetic Sensors—Development Trends and Applications*; Asfour, A., Ed.; IntechOpen: London, UK, 2017; pp. 39–62.

26. Crozier, R.; Mueller, M. A new MATLAB and octave interface to a popular magnetics finite element code. In Proceedings of the 22nd International Conference Electrical Machines (ICEM), Ecublens, Switzerland, 4–7 September 2016; pp. 1251–1256.

27. Lago-Cachón, D.; Rivas, M.; Martínez-García, J.C.; García, J.A. Cu impedance-based detection of superparamagnetic nanoparticles. *Nanotechnology* **2013**, *24*, 245501. [CrossRef] [PubMed]

28. Moyano, A.; Salvador, M.; Martinez-Garcia, J.C.; Socoliuc, V.; Vekas, L.; Peddis, D.; Alvarez, M.A.; Fernandez, M.; Rivas, M.; Blanco-Lopez, M.C. Magnetic immunochromatographic test for histamine detection in wine. *Anal. Bioanal. Chem.* **2019**, *25*, 6615–6624. [CrossRef] [PubMed]

29. De Cos, D.; Lete, N.; Fdez-Gubieda, M.L.; Garcia-Arribas, A. Study of the influence of sensor permeability in the detection of a single magnetotactic bacterium. *J. Magn. Magn. Mater.* **2020**, *500*, 166346. [CrossRef]

30. Fodil, K.; Denoual, M.; Dolabdijan, C.; Treizebre, A.; Senez, V. Model calculation of the magnetic induction generated by magnetic nanoparticles flowing into a microfluidic system: Performance analysis of the detection. *IEEE Trans. Magn.* **2014**, *50*, 5300108. [CrossRef]

 *sensors*

Article

Specific Loss Power of Co/Li/Zn-Mixed Ferrite Powders for Magnetic Hyperthermia

Gabriele Barrera [1,*], Marco Coisson [1], Federica Celegato [1], Luca Martino [1], Priyanka Tiwari [2,3], Roshni Verma [2], Shashank N. Kane [2], Frédéric Mazaleyrat [4] and Paola Tiberto [1]

[1] Nanoscience and Materials Division, Istituto Nazionale di Ricerca Metrologica (INRiM), Strada delle Cacce 91, I-10135 Torino, Italy; m.coisson@inrim.it (M.C.); f.celegato@inrim.it (F.C.); l.martino@inrim.it (L.M.); p.tiberto@inrim.it (P.T.)
[2] Magnetic Materials Laboratory, School of Physics, Devi Ahilya University, Khandwa road Campus, Indore 452001, India; priyanka.tiwari91092@gmail.com (P.T.); roshnikedar@gmail.com (R.V.); kane_sn@yahoo.com (S.N.K.)
[3] Department of Physics, Prestige Institute of Engineering Management and Research, Indore 452010, India
[4] Laboratory of Systems & Applications of Information & Energy Technologies (SATIE), ENS University Paris-Saclay, CNRS 8029, 61 Av. du Pdt. Wilson, F-94230 Cachan, France; Frederic.MAZALEYRAT@ens-cachan.fr
* Correspondence: g.barrera@inrim.it

Received: 12 March 2020; Accepted: 9 April 2020; Published: 10 April 2020

Abstract: An important research effort on the design of the magnetic particles is increasingly required to optimize the heat generation in biomedical applications, such as magnetic hyperthermia and heat-assisted drug release, considering the severe restrictions for the human body's exposure to an alternating magnetic field. Magnetic nanoparticles, considered in a broad sense as passive sensors, show the ability to detect an alternating magnetic field and to transduce it into a localized increase of temperature. In this context, the high biocompatibility, easy synthesis procedure and easily tunable magnetic properties of ferrite powders make them ideal candidates. In particular, the tailoring of their chemical composition and cation distribution allows the control of their magnetic properties, tuning them towards the strict demands of these heat-assisted biomedical applications. In this work, $Co_{0.76}Zn_{0.24}Fe_2O_4$, $Li_{0.375}Zn_{0.25}Fe_{2.375}O_4$ and $ZnFe_2O_4$ mixed-structure ferrite powders were synthesized in a 'dry gel' form by a sol-gel auto-combustion method. Their microstructural properties and cation distribution were obtained by X-ray diffraction characterization. Static and dynamic magnetic measurements were performed revealing the connection between the cation distribution and magnetic behavior. Particular attention was focused on the effect of Co^{2+} and Li^+ ions on the magnetic properties at a magnetic field amplitude and the frequency values according to the practical demands of heat-assisted biomedical applications. In this context, the specific loss power (SLP) values were evaluated by ac-hysteresis losses and thermometric measurements at selected values of the dynamic magnetic fields.

Keywords: magnetic hyperthermia; specific loss power; magnetic mixed ferrites; hysteresis losses; thermometric measurements

---

1. Introduction

Nanotechnology addressed to a nanoscale design of materials is one of the utmost researched topics in the present century, involving disciplines like engineering, physics, chemistry and biology, concerning different application areas such as electronics, telecommunications, energy harvesting, sensors and biomedicine [1–7].

Although magnetic nanoparticles have been extensively studied in recent decades, they resulted in exciting materials to be used in these application areas due to their considerably size-dependent chemical and physical properties [8–10]. In particular, in the biomedical area, the tuning of the structure, size and composition of particles has led to the development of different applications such as magnetic biosensors, magnetic resonance imaging (MRI), drug-delivery and magnetic hyperthermia [6,11–20].

When exposed to an alternating magnetic field, magnetic particles, considered in a broad sense as passive sensors, are able to detect and transduce it in a controlled and localized release of heat; this ability has promoted the use of these materials for advanced therapeutic applications such as magnetic hyperthermia and heat-assisted drug release [13,21–23]. The physical mechanism at the base of heat generation has been recently identified to be mainly the magnetic hysteresis losses [24]. The efficiency of the heat generation, estimated by the specific loss power (SLP) value, depends on several parameters; some of these can be identified as "external", such as the intensity and the frequency of the applied magnetic field and the liquid medium properties, whereas others, identified as "internal", depend on the intrinsic properties of the magnetic particles such as composition, size, shape and magnetic state [25,26].

Because of the strict restrictions required on the applied alternating magnetic field parameters for the human body's exposure [27–29], a huge research effort should be primarily focused on the design of the intrinsic properties of magnetic particles in order to optimize the efficiency of the heat release according to practical demands [30,31].

In this context, spinel ferrite powders attract extraordinary attention because of their high biocompatibility, easy synthesis procedure, physical and chemical stability and easily tunable magnetic properties [32–37]. The general formula of ferrite is $MeFe_2O_4$ where Me represents a divalent metal ion (e.g., Fe^{2+}, Co^{2+}, Ni^{2+}, Zn^{2+}, etc.) or a combination of ions with the average valence of two (e.g., Li^+ and Fe^{3+} in lithium ferrite, etc.) [33]. Moreover, combinations of these ions are also possible, obtaining mixed-structure ferrites with different compositions [33,38].

Generally, in the biomedical area, ferrite particles based on magnetic Co^{2+} ions are proposed as promising heat generators due to their strong magnetocrystalline anisotropy and moderate magnetization [39–41]. Instead, Li^+ ions, combined in magnetic ($Li^+_{0.5}Fe^{3+}_{0.5}$) species in ferrite structures, have attracted attention because of their low toxicity [42,43]. On the other hand, the non-magnetic Zn^{2+} ion is usually used as a partial substitutional of magnetic divalent ions or species in order to finely tune the magnetic properties of the ferrite particles [38], especially the saturation magnetization [33].

Together with the chemical composition changes, the manipulation of the cation distribution on octahedral and tetrahedral sites represents another suitable strategy to control the magnetic behavior of ferrites due to the strong connection between the spinel structure and its magnetism [44–47].

The biocompatibility evaluation of ferrite particles represents a preliminary and fundamental step towards their use in biomedical applications. The variety of ions in the ferrite composition represents one of the several parameters that influence the viability of the cells [48,49]; e.g., Co-ferrite is less biocompatible than Fe_3O_4 and Mn-ferrite [50]. However, different coating materials such as polymer or surfactants can be used as protective layers minimizing the direct exposure of the less biocompatible ions present on the ferrite surface to the biological environment [39,51].

In the present work, $Co_{0.76}Zn_{0.24}Fe_2O_4$, $Li_{0.375}Zn_{0.25}Fe_{2.375}O_4$ and $ZnFe_2O_4$ mixed-structure ferrites were synthesized in a 'dry gel' form by a sol-gel auto-combustion method. Here, X-ray diffraction (XRD) was exploited to calculate the cation distribution. Static and dynamic magnetic characterizations were performed to study the connection between the cation distribution and magnetic behavior. Particular attention was devoted to clarifying the role of Co^{2+} and Li^+ ions on the magnetic behavior of ferrites at a magnetic field amplitude and the frequency values appropriate to foresee heat-assisted biomedical applications. Moreover, the hysteresis losses and thermometric measurements at selected values of ac-magnetic fields were performed to evaluate the specific loss power (SLP) of the samples.

2. Materials and Methods

$Li_{0.375}Zn_{0.25}Fe_{2.375}O_4$, $Co_{0.76}Zn_{0.24}Fe_2O_4$ and $ZnFe_2O_4$ powder ferrite samples were synthesized by sol-gel auto-combustion method [52]. In summary, the powders were synthesized via utilizing AR grade citrate-nitrate/acetate precursors ($Zn(NO_3)_2.6H_2O$, Ferric nitrate—$Fe(NO_3)_3.9H_2O$, Lithium acetate—$CH_3.COOLi.2H_2O$, cobalt nitrate—$Co(NO_3)2\ 6H_2O$ and citric acid—$C_6H_8O_7$), were mixed in the stoichiometric ratio. Citric acid was used as the 'fuel' and the ratio of metal salts to fuel was taken as 1:1. In the beginning, the citric acid acted as a chelating agent for the metal ions of varying ionic sizes, which helped in preventing their selective precipitation to maintain compositional homogeneity among the constituents. Subsequently, it also served as a fuel in the combustion reaction. To start the synthesis process, all the precursor materials for the desired composition were dissolved in deionized water in a beaker under constant stirring, to get a homogeneous solution and the pH was maintained at 7 by adding ammonia solution. Then, the solution was heated at ~110 °C for 1 h in air till a fluffy powder was formed called 'dry gel' or 'as-burnt powder', which was ground to get a fine powder. After the synthesis, the $Li_{0.375}Zn_{0.25}Fe_{2.375}O_4$ and $ZnFe_2O_4$ samples were annealed at 450 °C for 3 h; instead, no post-synthesis heat treatments were performed on the $Co_{0.76}Zn_{0.24}Fe_2O_4$ sample.

The Zn concentration (x $\approx$ 0.25) and the resulting composition of LiZn- and CoZn-ferrites were chosen in order to induce an increase in the saturation magnetization compared to the corresponding simple ferrites (CoFeO and LiFeO) [33,53], taking into account the practical requests to optimize the heating efficiency. A higher concentration of Zn ions (x > 0.5) should be avoided as it lowers the magnetization saturation. [33,53].

The structural properties of powder samples were analyzed by X-ray diffraction (XRD) using Cu-K_α radiation (wavelength 'λ' = 0.1541 nm) in θ–2θ configuration (step size of 0.02°), equipped with a fast counting Bruker LynxEye detector, with Silicon strip technology. Rietveld refinement was performed by MAUD (material analysis using diffraction) software [54]. XRD data were analyzed to obtain structural parameters: experimental ($a_{exp.}$) lattice parameters, X-ray density (ρ_{xrd}) and mean grain diameter ($<D_{xrd}>$). The distribution of cations on tetrahedral and octahedral sites in the studied samples was determined by analyzing the XRD pattern, employing the Bertaut method [55], as was also reported in earlier reports [56–58]. XRD intensity depends on the atomic position in the spinel unit cell whereas the XRD peak position relies on the size and the shape of the unit cell. Bertaut's method utilizes the following pair of reflections: (400), (422) and (220), (400), according to the expression:

$$\frac{I^{obs}_{hkl}}{I^{obs}_{h'k'l'}} = \frac{I^{cal}_{hkl}}{I^{cal}_{h'k'l'}} \tag{1}$$

where $I_{hkl}{}^{obs}$ and $I_{hkl}{}^{cal}$ are respectively the observed and estimated intensities for the reflection (hkl). These ratios were evaluated for the numerous groupings of cationic distribution at tetrahedral and octahedral sites as described in [56]. The best distribution of cations was taken among the sites for which theoretical and experimental lattice parameters agreed clearly.

Room-temperature-static-magnetization curves were measured by the means of a vibrating sample magnetometer (VSM), operating in the magnetic field range −1200 < H < 1200 kA/m. The dc-hysteresis loops were measured at selected vertex fields in the interval 0–1200 kA/m. From the major loops, the coercive field (H_c), the magnetic remanence (μ_0M_r) and the saturation magnetization (μ_0M_s) were evaluated; specifically, the latter was determined by fitting the high-field portion of dc-hysteresis curves with the standard expression $\mu_0M = \mu_0M_s - a/H$, representing the first-order approximation of the series expansion that describes the law of approach to saturation [59].

Room-temperature dynamic hysteresis loops were measured by the means of a custom-built B-H loop tracer [60] operating with the ac-magnetic field amplitude in the range 8–42 kA/m at the fixed frequency of ~69 kHz; the ac-hysteresis loops were measured at selected vertex fields in the allowable range. The area enclosed by ac-hysteresis loops was calculated in order to estimate the specific loss power by the means of the hysteresis losses (SLP).

Thermometric measurements were performed by an ad hoc-developed hyperthermia setup described in detail elsewhere [60]. In summary, the magnetic particle suspension at desired concentration was placed in the center of a copper coil with a diameter of about 5 cm which generated an electromagnetic field with a frequency of 100 kHz and an intensity up to 47.7 kA/m. The geometry of the set-up guaranteed a homogenous field through the entire sample and the used field frequency and intensity fell within the general safety [27,28]. The magnetic particle suspension was prepared by dispersing the $Li_{0.375}Zn_{0.25}Fe_{2.375}O_4$ powder in a liquid medium, consisting of sodium citrate tribasic dispersed in deionized water (0.2 g/L), obtaining a magnetic solution concentration of about 27 mg/mL. Sodium citrate tribasic favors the stabilization of ferrite particles in deionized water due to the adsorption of citrate anions onto their surface resulting in a stable and well dispersed magnetic solution [61]. These liquid suspensions were adequate to perform thermometric measurements but did not present conditions for in vitro and in vivo biomedical applications. The thermodynamic conditions of the experiment were fully modeled to obtain a direct measurement of the SLP of the magnetic powders by taking into account the heat exchange with the surrounding environment in non-adiabatic conditions and the parasitic heating of the water [60].

3. Results and Discussion

3.1. Structural Properties

Rietveld-refined X-ray diffraction (XRD) patterns of the studied samples validating the formation of the nanocrystalline mixed cubic spinel structure are presented in Figure 1. For the sample of $ZnFe_2O_4$ (Figure 1c), small traces of Fe_2O_3 were also observed and can be ascribed to a partial decomposition of the ferrite phase [62].

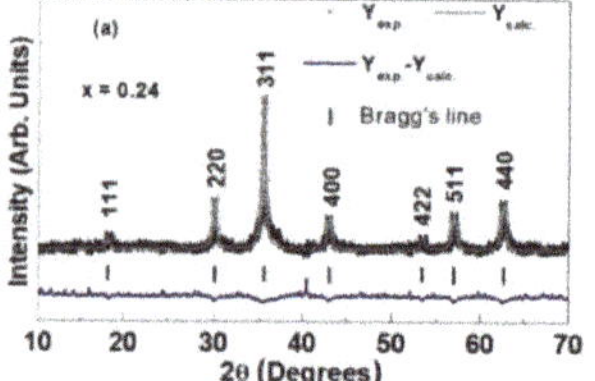
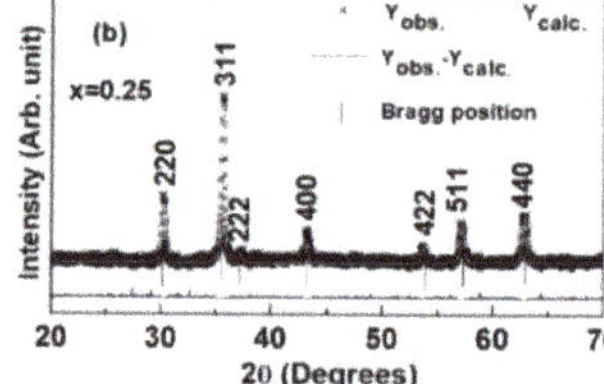
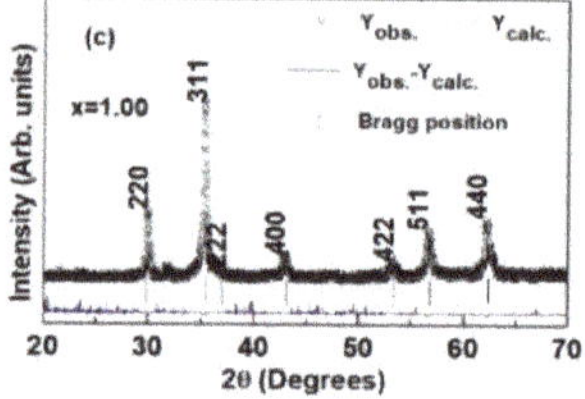

Figure 1. Rietveld-refined XRD patterns of: (**a**) dry gel $Co_{0.76}Zn_{0.24}Fe_2O_4$, (**b**) $Li_{0.375}Zn_{0.25}Fe_{2.375}O_4$ annealed at 450 °C/3 h, (**c**) $ZnFe_2O_4$ annealed at 450 °C/3 h.

The experimental lattice parameter (a_{exp}) values of the ferrite phases obtained from the Rietveld refinement of the XRD data are given in Table 1; the value of the Zn-ferrite sample results is higher than the ones of the LiZn-ferrite and CoZn-ferrite samples. This observed increase was due to the replacement of the Co^{2+} and Li^+ ions characterized by an ionic radius of 0.78 and 0.70 Å, respectively, with the bigger Zn^{2+} ions (0.83 Å) [63–65], in agreement with Vegard's law [66]. Moreover, the X-ray density (ρ_{XRD}) for each sample was calculated by using the lattice parameter (a_{exp}) and the standard formula [67]; the obtained values, ranging in the typical interval of this kind of ferrites [33,68], are given in Table 1.

Table 1. Experimental lattice parameters (a_{exp}); X-ray density (ρ_{XRD}); grain mean diameter $<D_{XRD}>$ and cation distribution.

Sample	a_{exp} (nm)	ρ_{XRD} (g/cm³)	$<D_{XRD}>$ (nm)	Cation Distribution
$Co_{0.76}Zn_{0.24}Fe_2O_4$	0.8391	5.3	32	$(Co_{0.00}Zn_{0.10}Fe_{0.90})$ $[Co_{0.76}Zn_{0.14}Fe_{1.10}]O_4$
$Li_{0.375}Zn_{0.25}Fe_{2.375}O_4$	0.8365	4.9	38	$(Li_{0.05}Zn_{0.10}Fe_{0.85})$ $[Li_{0.325}Zn_{0.15}Fe_{1.525}]O_4$
$ZnFe_2O_4$	0.8435	5.3	35	$(Zn_{0.09}Fe_{0.91})$ $[Zn_{0.91}Fe_{1.09}]O_4$

The mean diameter $<D_{XRD}>$ values of the grains were obtained from the XRD data analysis by Scherrer's equation [69] for all the samples (Table 1) and they were in the interval 32–38 nm, indicating the formation of nanocrystallites. However, the reactions used to synthesize the particles could induce a non-negligible degree of agglomeration, which could result in polycrystalline aggregates forming particles with diameters larger than $<D_{XRD}>$ [44].

These small values of $<D_{XRD}>$ revealed a significant advantage of the auto-combustion synthesis compared to the more conventional ceramic methods in order to produce ferrite particles: the lower temperature and shorter time used in the auto-combustions synthesis resulted in the grains' smaller diameter with a greater surface area [70,71], preventing the particles from coarsening and aggregating, which is promoted by the very high temperature (T > 1000 °C) typically used in the ceramic methods.

In addition, the XRD spectra were also analyzed by the Bertaut method [55] in order to obtain the cation distribution. In particular, the cation distributions of the studied samples are given in Table 1, where the ions on the tetrahedral site (site A) are given in parentheses and the ions on the octahedral sites (sites B) between square brackets. The Zn^{2+} ions in Zn-ferrite were mainly located on the B site, whereas the CoZn-ferrite and LiZn-ferrite samples showed only a slight preference of the Zn^{2+} ions to occupy the octahedral (B) sites with respect to the tetrahedral (A) one. These observed deviations, with respect to the general preference of the Zn^{2+} ions to occupy the tetrahedral (A) site, clearly proved a non-equilibrium cation distribution in the samples. This effect was ascribed to the sol-gel auto-combustion synthesis method at a low temperature (<110 °C) and to the low efficiency of the post-synthesis heat treatment (for sample Zn-ferrite and LiZn-ferrite) which did not effectively induce a diffusion of the Zn^{2+} ions on the tetrahedral sites towards a distribution closer to the equilibrium. Instead, in the CoZn-ferrite and LiZn-ferrite samples, the divalent Co^{2+} metal ions and the divalent combination of two the metal ions $[Li^+_{0.5}Fe^{3+}_{0.5}]$ [68,72] were almost totally located, as their preference, on B sites.

The cation distribution reported as the general formula $\left(Me^{II}_{\delta}Fe^{III}_{1-\delta}\right)\left[Me^{II}_{1-\delta}Fe^{III}_{1+\delta}\right]O_4$ revealed the inversion degree of all the studied samples ($\delta = 0.10, 0.15, 0.09$ for the CoZn, LiZn and Zn-ferrite samples, respectively), indicating an intermediate configuration of their structure between the completely inverse spinel structure ($\delta = 0$) and the totally random distribution one ($\delta = 0.33$).

The intermediate and tunable values of the inversion degree and its effect on the magnetic properties (see the following sections) are advantages ascribed to the auto-combustion synthesis, in which the low temperature and the fast cooling rate hinder the diffusion of the metal ions towards the equilibrium that it is typically obtained with the more conventional high-temperature ceramic methods.

3.2. DC-Magnetic Properties

An enlargement of the room-temperature dc-hysteresis loops of all the samples taken at the maximum magnetic field of 1200 kA/m is shown in Figure 2.

All curves display the typical magnetic hysteretic behavior resulting from the ferrimagnetic ordering of spinel ferrite structure [73,74], confirming the blocked state of the ferrite particles as expected by the $<D_{XRD}>$ values of the grains obtained from the XRD data analysis. The observed coercive field (H_c), saturation magnetization ($\mu_0 M_s$) and the magnetic remanence ($\mu_0 M_r$) are listed in Table 2.

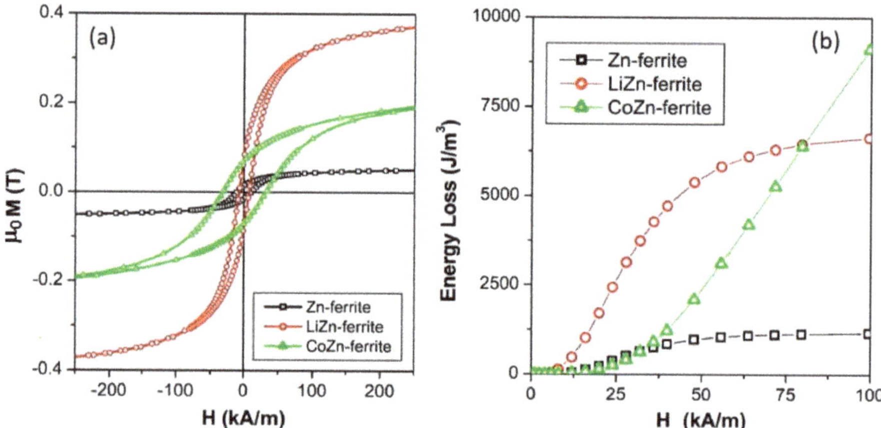

Figure 2. (**a**) Room-temperature major dc-hysteresis loops of all the studied samples; (**b**) dc-hysteresis loops areas as a function of the vertex field for all the studied ferrites.

Table 2. Saturation magnetization ($\mu_0 M_s$), coercive field (H_c), magnetic remanence ($\mu_0 M_r$), theoretical saturation magnetization ($\mu_0 M_s^{th}$) at 0 K and the area enclosed by the major dc-hysteresis loops for all the samples.

Sample	$\mu_0\mathbf{M_s}$	$\mathbf{H_c}$	$\mu_0\mathbf{M_r}$	$\mu_0\mathbf{M_s}^{th}$	Area
	(T)	(kA/m)	(T)	(T)	(J/m^3)
$Co_{0.76}Zn_{0.24}Fe_2O_4$	0.255	31.90	0.071	0.52	21709
$Li_{0.375}Zn_{0.25}Fe_{2.375}O_4$	0.415	7.37	0.078	0.54	7826
$ZnFe_2O_4$	0.066	10.65	0.015	0.14	1467

The complete replacement of the magnetic Co^{2+} ions with the non-magnetic Zn^{2+} ions induced a reduction of H_c from 31.90 (CoZn-ferrite) to 10.65 (Zn-ferrite) kA/m. The higher value of H_c of CoZn-ferrite sample was connected to the high anisotropy of Co^{2+} ions characterized by a remarkable spin-orbit coupling [36,75]. On the other hand, the replacement of the magnetic [$Li^+_{0.5}Fe^{3+}_{0.5}$] species with the non-magnetic Zn^{2+} ions promoted a slight increase of H_c from 7.37 (LiZn-ferrite) to 10.65 kA/m (Zn-ferrite) and also a marked reduction of the $\mu_0 M_s$ values from 0.415 T (LiZn-ferrite) to 0.066 T (Zn-ferrite). This $\mu_0 M_s$ reduction can be ascribed to the different distribution of the Fe^{3+} ions in the spinel structure (see cation distribution in Table 1). In particular, in the LiZn sample, the Fe^{3+} ions were preferentially located on the octahedral sites, leading to a higher net magnetic moment than the one of the Zn sample in which a quasi-balanced distribution of Fe^{3+} ions occurred.

The experimental $\mu_0 M_s$ values can be compared to the theoretical magnetization ($\mu_0 M_s^{th}$) values (see Table 2) calculated in accordance to Néel's two-sublattice model of ferrimagnets [76], where the magnetic moments of the ions (Fe^{3+} = 5 μ_B, Co^{2+} = 3 μ_B and Li^+ = Zn^{2+} = 0 μ_B) on the tetrahedral and octahedral sites are considered perfectly anti-parallel, totally neglecting any temperature effect and spin disorder. The $\mu_0 M_s^{th}$ values for the LiZn-ferrite and CoZn-ferrite samples were very similar; therefore, the experimental $\mu_0 M_s$ values are also expected to be comparable. Instead, a marked difference was measured (Table 2). This evidence clearly proves that the magnetic moments in these two ferrite samples were not perfectly antiparallel, but rather were characterized by a spin canting resulting in a non-collinear arrangement in the two-sublattices [77], especially in the CoZn-ferrite samples. The three-lattice model, suggested by Yafet and Kittel [77], confirms this hypothesis by extrapolating the canting angle values from the XRD data: 38°, 30° and 24° for the CoZn-ferrite, LiZn-ferrite, and the Zn-ferrite, respectively. Of course, the observed reduction of $\mu_0 M_s$ can also be accentuated by the probable presence of a spin disorder on the ferrite surface that induced a magnetically dead layer [78,79].

The calculated values of the area (see Table 2) enclosed by the dc-hysteresis loops were proportional to the energy lost as heat by the samples in one complete major loop (hysteresis losses). It can be noted that the mixing of magnetic divalent ions (Co^{2+}) or species [$Li^+_{0.5}Fe^{3+}_{0.5}$] with non-magnetic Zn^{2+} ions increased the area enclosed by the major loop with respect to the Zn-ferrite structure. This effect was particularly efficient in the CoZn-ferrite sample due to the high anisotropy of Co^{2+} ions.

Minor dc-hysteresis loops were measured for all the samples by applying selected vertex fields lower than the saturation field of the CoZn sample. The values of the area enclosed by these minor dc-loops were calculated in order to evaluate the dc-hysteresis losses as a function of selected the vertex fields (H_v), see Figure 2b. All samples showed a non-linear dependence of the dc-hysteresis losses on the vertex field amplitude, which can be well described by a third-order power law in the limit of the small applied field [59]. Up to the field value of about 75 kA/m, the mutual relationship of the dc-hysteresis losses intensity of the three samples was markedly changed with respect to the dc-hysteresis losses evaluated from major dc-hysteresis loops (Table 2). In particular, in the vertex field interval $0 < H_v < 75$ kA/m, the LiZn-ferrite sample showed the highest hysteresis losses values whereas those of the CoZn-ferrite sample were very low because the high anisotropy of Co^{2+} ions was not completely overcome and the magnetization described narrow minor loops. When $H_v > 75$ kA/m, the area enclosed by CoZn samples becomes the highest among the studied samples restoring the mutual relationship of the dc-hysteresis losses intensity evaluated from major dc-hysteresis loops.

3.3. Ac-Measurements and SLP Evaluation

Magnetic hyperthermia therapy has to comply with a variety of biological and technical constraints, among which emerges the limit for the product of magnetic field amplitude and frequency ($H \times f$) in relation to the induced current loop diameter (D) to avoid non-specific heating in healthy areas due to Eddy currents and to avoid the stimulation of cardiac muscles and nerves [80,81]. This limit was initially proposed by Atkinson et al., as $H \times f < 4.85 \times 10^8$ $Am^{-1}s^{-1}$ for a loop diameter of about 30 cm [27]; further experiments with a smaller diameter of the exposed body region have increased the criterion up to $H \times f < 5.0 \times 10^9$ $Am^{-1}s^{-1}$ [28].

The range between these two criteria seems to be currently the most convenient and most used in several in vivo trials [82], at least until new studies about the biological safety of the alternating magnetic field are carried out.

In addition to this biological limitation, some engineering difficulties combined with increasing costs arise with the aim of concurrently increasing the field frequency and amplitudes in the wide space (several centimeters) required for hyperthermia treatment.

In this work, the ac-characterization was performed at the field frequency of 69 kHz and at the maximum applied magnetic field intensity, limited to 42 kA/m in order to fall within both general safety [27,28] and our technological limits. A selection of ac-hysteresis loops acquired at selected vertex fields for all samples is shown in Figure S1 of the Supplementary Materials.

Dynamic ac-hysteresis loops ($f = 69$ kHz and $H_v = 37$ kA/m) of all the studied powder ferrite samples are shown in Figure 3a.

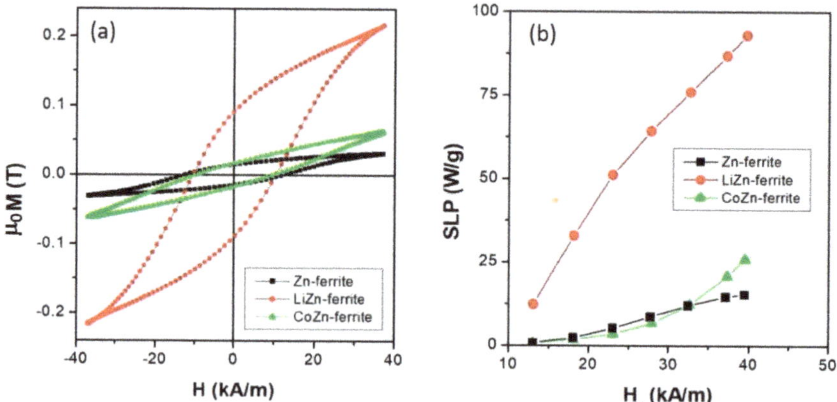

Figure 3. (**a**) Room-temperature minor ac-hysteresis loops of all the studied samples (f = 69 kHz and H_v = 37 kA/m); (**b**) specific loss power (SLP) values for all the samples as a function of the vertex field obtained by the ac-hysteresis loops areas.

In all the samples, the magnetization did not reach complete saturation, leading to minor loops. The area enclosed by the hysteresis loops represented the irreversible work dissipated in the surrounding as thermal energy profitably usable for magnetic hyperthermia [60,83–85].

The specific loss power is defined as the power released in the form of heat by powders submitted to electromagnetic field radiation, normalized to the mass of the solid component of the sample. In our case, the SLP was calculated from the area enclosed by the ac-hysteresis loops by the means of the following integral of the dynamic magnetization versus the applied field strength:

$$SLP = \frac{f}{c} \oint \mu_0 M(t) dH(t) \tag{2}$$

where f is the field frequency and c the weight concentration of ferrite powder. The integration was done over one period of the oscillating magnetic field.

The SLP dependence on the vertex field intensity H_v of all the studied samples is shown in Figure 3b. The evaluation of this parameter represented one of the primary criteria to determine the suitability of the ferrite sample for hyperthermia applications.

In the entire investigated field range, the LiZn-ferrite sample displayed the highest SLP values indicating how the low-magnetic anisotropy divalent species [$Li^+_{0.5}Fe^{3+}_{0.5}$] favor the heat released by hysteresis losses with respect to the high-magnetic anisotropic divalent Co^{2+} ions; in particular, the SLP value of the LiZn-ferrite sample at H_v = 42 kA/m is 3.6 times bigger than the one of CoZn-ferrite sample at the same vertex field. As observed from the dc-characterization reported in Figure 2b, only a further increase of H_v (exceeding our technological limit) can reduce the gap between the SLP values produced by two samples leading to the promotion of the CoZn-ferrite sample. Moreover, the total replacement of the magnetic divalent ions Co^{2+} or species [$Li^+_{0.5}Fe^{3+}_{0.5}$] with non-magnetic Zn^{2+} ions hinders the heat production by hysteresis losses, resulting in low SLP values in the investigated field range with a tendency to reach a saturation value of about 15 W/g.

3.4. Thermometric Measurements and SLP Evaluation

Thermometric measurements were conducted in order to evaluate the ability of ferrite powders to heat the magnetic solution, in which they were dispersed, under different applied field intensities. Only the LiZn-ferrite composition was examined by thermometric measurements because it resulted in, by the ac-loops characterization, the most promising sample among those studied for hyperthermia therapy. In particular, the time dependence of the temperature of the magnetic solution containing the LiZn-ferrite powder (concentration of about 27 mg/mL) under an applied field for 40 kA/m at 100 kHz is shown in Figure 4.

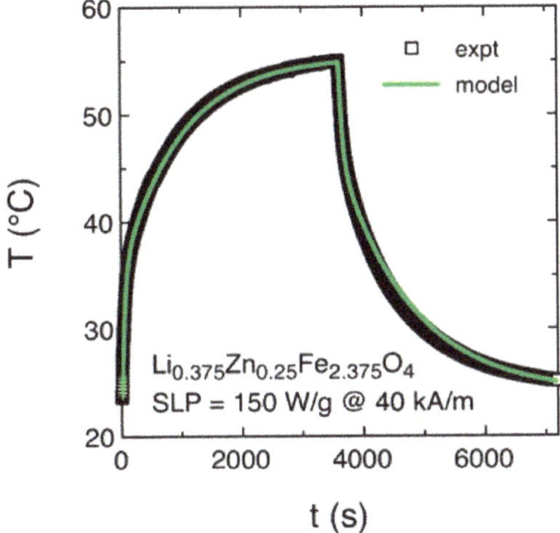

Figure 4. Time dependence of the temperature of the magnetic solution containing LiZn-ferrite powder under an applied field of 40 kA/m at 100 kHz. Black symbols: experimental data. Green line: best fit the by theoretical model.

The reported curve is taken as representative of all the measurements done at different applied field values from 23.9 to 47.7 kA/m. At $t = 0$ s, the radiofrequency field was switched on and the temperature of the magnetic solution increased towards the equilibrium temperature. After one hour, the radiofrequency field was switched off and the magnetic solution cooled down to room temperature. As a result, the thermometric SLP values were extrapolated by fitting the whole time evolution of the temperature of the magnetic solution by a mathematical model that takes into account the non-adiabatic condition of the exploited hyperthermia setup [60]. In Figure 4, the black symbols are the experimental data, whereas the green line is the theoretical best fit. The model based on Newton's cooling law [86] evaluates the exchange of heat among the various components of the setup and/or the surrounding environment induced by their temperature difference. In particular, the heat released by the magnetic particles excited by an r.f electromagnetic field is all transferred to the liquid medium in which they are dispersed. Subsequently, the heat is transferred to all the experimental components and eventually to the surrounding environment. An accurate calibration procedure allows to determine the time constant of the heat exchange mechanism among the various experimental components and the other physical quantities appearing in the mathematical model. The only free parameter of the fit procedure is the power (P) released by the ferrite powders that it is adjusted to reproduce the whole experimental curve providing a direct estimation of the thermometric SLP of the samples:

$$SLP = \frac{P}{m} \tag{3}$$

where m is the mass of ferrite powders in the liquid solution.

A summary of the SLP values determined with the two techniques for LiZn-ferrite samples is plotted in Figure 5. The SLP values calculated from thermometric measurements (empty red dots in Figure 5) turned out to be larger than the ones obtained by dynamic hysteresis loops (full red dots in Figure 5).

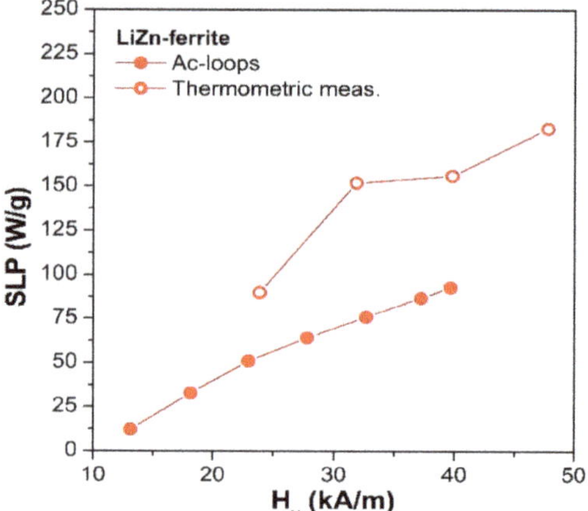

Figure 5. SLP values for the LiZn sample as a function of the vertex field obtained by: ac-hysteresis loops areas (full red dots) at the operation frequency of 69 kHz and the thermometric measurements (empty red dots) at the operation frequency of 100 kHz.

This effect was mainly due to the higher operating frequency of the thermometric setup (100 kHz) with respect to the B-H loop tracer (69 kHz) and it also included the contribution to the SLP of the Brown relaxation process that took place only in the magnetic liquid solution and not in the dried sample used for dynamic hysteresis loops measurements [24,87]. In fact, the Brown relaxation mechanism is associated with the physical rotation of the whole particle in the fluid generating heat due to the viscous friction between the rotating particles and the surrounding liquid medium [24,87,88]. However, recent studies, both in vivo and ex vivo, have demonstrated that the particles are generally immobilized when directly injected into the tumor tissues highlighting that the Brown process is largely suppressed during hyperthermia treatments [89–91]. Another significant aspect to be taken into consideration is the effect of the strong magnetic interaction that is generated by the agglomerations of particles in dried samples. In particular, the formation of agglomerates can entail an alteration of the hysteresis loop area and consequently influence the SLP value [92–94].

SLP values obtained for the LiZn-ferrite sample (SLP = 85–180 W/g) as reported in Figure 5 are in the same order of magnitude as the ones already reported in the literature. As examples, Mallik et al. [42] report SLP = 334 W/g (H = 33.5 mT, f = 290 kHz) for $Li_{0.31}Zn_{0.38}Fe_{2.31}O_4$ particles; Dalal et al. [95] report SLP = 143–166 W/g (H = 33.5 mT, f = 290 kHz) for $Li_{0.35}Zn_{0.3}Co_{0.05}Fe_{2.3}O_4$ particles embedded in carbon nanotubes. It should be noted that the experimental results are hardly comparable as they refer to different applied fields and frequencies and strongly depend on the intrinsic properties of particles such as size, shape and composition. This comparison shows that Li-based ferrites can be considered an efficient alternative to more conventional iron-oxides such as maghemite (SLP = 106 W/g [96]) and magnetite [60] and some other ferrite compositions (SLP = 300 W/g for Gd-ferrite [97], SLP = 73 W/g for Ba-ferrite [98] and SLP = 58 W/g for Sr-ferrite [99]); however, the heat efficiency of Li-based ferrites remains much lower than Mn- and MnCo-based ferrites (SLP = 3024 W/g [100]) and magnetosomes (SLP = 960 W/g [101]).

4. Conclusions

$Co_{0.76}Zn_{0.24}Fe_2O_4$, $Li_{0.375}Zn_{0.25}Fe_{2.375}O_4$ and $ZnFe_2O_4$ mixed-structure ferrite powders were synthesized by a sol-gel auto-combustion method. XRD spectra analysis revealed an out-of-equilibrium cations distribution ascribed to the low efficiency of the synthesis method which did not induce an effective diffusion of the Zn^{2+} ions towards the equilibrium positions. This out-of-equilibrium cation distribution strongly influenced the static and dynamic magnetic properties of the ferrites.

Dc-major loops clearly indicated that the suitable mixing of non-magnetic Zn^{2+} ions with magnetic divalent ions (Co^{2+}) or species [$Li^{+}_{0.5}Fe^{3+}_{0.5}$] is a tool to tune the magnetic properties of the ferrite particles. In particular, the high anisotropy of Co^{2+} ions enhanced the coercive field, whereas the divalent species [$Li^{+}_{0.5}Fe^{3+}_{0.5}$] increased the saturation magnetization.

The magnetic energy dissipated as thermal energy profitably usable in hyperthermia was evaluated by the means of ac-hysteresis loops and thermometric measurements. Among the studied samples, the mixed LiZn-ferrite structure matches in a better way the practical requirements of heat-assisted applications than those of the mixed ferrite structure constituted by the highly anisotropic Co^{2+} ions. In the entire investigated field range, the LiZn-ferrite sample displayed the highest SLP values indicating how the low anisotropic divalent species [$Li^{+}_{0.5}Fe^{3+}_{0.5}$] favor the heat release by hysteresis losses with respect to the high anisotropic divalent Co^{2+} ions.

In conclusion, the SLP evaluation and the structural and magnetic characterizations of ferrite samples in dry form or dispersed in the liquid solution were calculated. This represents a preliminary and important step towards the understanding of the physical properties for their perspective use in magnetic hyperthermia and heat-assisted biomedical applications. Obviously, for prospective in vitro and in vivo studies, several other aspects should be carefully taken into account such as the size and charge of the NPs and their degree of aggregation in the magnetic solution.

Supplementary Materials: The following are available online at http://www.mdpi.com/1424-8220/20/7/2151/s1, Figure S1: A selection of the ac-hysteresis loops acquired at the selected vertex fields for all samples.

Author Contributions: Conceptualization, P.T. (Paola Tiberto), M.C. F.M. and S.N.K.; preparation of the samples, P.T. (Priyanka Tiwari) and R.V.; XRD characterization and structural data analysis, P.T. (Priyanka Tiwari), R.V. and S.N.K.; DC-magnetic measurements and data analysis, G.B., M.C., F.C. and F.M.; AC-magnetic measurements and data analysis, G.B., M.C., F.C. and L.M.; SLP evaluation, G.B., M.C., F.C and L.M.; supervision, P.T. (Paola Tiberto), M.C. and S.N.K.; writing the manuscript, G.B., M.C. and S.N.K. All authors have read and agreed to the published version of the manuscript.

Funding: This research received no external funding.

Conflicts of Interest: The authors declare no conflict of interest.

References

1. Bhushan, B. *Springer Handbook of Nanotechnology*, 3rd ed.; Bhushan, B., Ed.; Springer: Heidelberg, Germany, 2010; ISBN 978-3-642-02524-2.

2. Hussein, A.K. Applications of nanotechnology in renewable energies—A comprehensive overview and understanding. *Renew. Sustain. Energy Rev.* **2014**, *42*, 460–476. [CrossRef]

3. *Bio-Inspired Nanotechnology*; Knecht, M.R.; Walsh, T.R. (Eds.) Springer: New York, NY, USA, 2014; ISBN 978-1-4614-9445-4.

4. Sposito, A.J.; Kurdekar, A.; Zhao, J.; Hewlett, I. Application of nanotechnology in biosensors for enhancing pathogen detection. *Wiley Interdiscip. Rev. Nanomed. Nanobiotechnol.* **2018**, *10*, e1512. [CrossRef] [PubMed]

5. Ye, W.; Xu, Y.; Zheng, L.; Zhang, Y.; Yang, M.; Sun, P. A nanoporous alumina membrane based electrochemical biosensor for histamine determination with biofunctionalized magnetic nanoparticles concentration and signal amplification. *Sensors* **2016**, *16*, 1767. [CrossRef] [PubMed]

6. Rocha-Santos, T.A.P. Sensors and biosensors based on magnetic nanoparticles. *TrAC Trends Anal. Chem.* **2014**, *62*, 28–36. [CrossRef]

7. Barrera, G.; Celegato, F.; Coisson, M.; Manzin, A.; Ferrarese Lupi, F.; Seguini, G.; Boarino, L.; Aprile, G.; Perego, M.; Tiberto, P. Magnetization switching in high-density magnetic nanodots by a fine-tune sputtering process on large area diblock copolymer mask. *Nanoscale* **2017**, *9*, 16981–16992. [CrossRef]

8. Sattler, K.D. *Handbook of Nanophysics. 3, Nanoparticles and Quantum Dots*; CRC Press: Boca Raton, FL, USA, 2011; ISBN 9781420075441.

9. Zeng, H.; Sun, S. Syntheses, Properties, and Potential Applications of Multicomponent Magnetic Nanoparticles. *Adv. Funct. Mater.* **2008**, *18*, 391–400. [CrossRef]

10. Khanna, L.; Gupta, G.; Tripathi, S.K. Effect of size and silica coating on structural, magnetic as well as cytotoxicity properties of copper ferrite nanoparticles. *Mater. Sci. Eng. C* **2018**, *97*, 552–566. [CrossRef]

11. Sun, C.; Lee, J.S.H.; Zhang, M. Magnetic nanoparticles in MR imaging and drug delivery. *Adv. Drug Deliv. Rev.* **2008**, *60*, 1252–1265. [CrossRef]

12. Hedayatnasab, Z.; Abnisa, F.; Daud, W.M.A.W. Review on magnetic nanoparticles for magnetic nanofluid hyperthermia application. *Mater. Des.* **2017**, *123*, 174–196. [CrossRef]

13. Thanh, N.T.K.T. *Magnetic Nanoparticles from Fabrication to Clinical Applications*; CRC Press: Boca Raton, FL, USA, 2012; Volume 54, ISBN 9781439869321.

14. Amiri, M.; Salavati-Niasari, M.; Pardakhty, A.; Ahmadi, M.; Akbari, A. Caffeine: A novel green precursor for synthesis of magnetic $CoFe_2O_4$ nanoparticles and pH-sensitive magnetic alginate beads for drug delivery. *Mater. Sci. Eng. C* **2017**, *76*, 1085–1093. [CrossRef]

15. Maksoud, M.I.A.A.; El-sayyad, G.S.; Ashour, A.H.; El-batal, A.I.; Abd-elmonem, M.S.; Hendawy, H.A.M.; Abdel-khalek, E.K.; Labib, S.; Abdeltwab, E.; El-okr, M.M. Synthesis and characterization of metals-substituted cobalt ferrite as antimicrobial agents and sensors for Anagrelide determination in biological samples. *Mater. Sci. Eng. C* **2018**, *92*, 644–656. [CrossRef] [PubMed]

16. Rodrigues, R.O.; Baldi, G.; Doumett, S.; Garcia-Hevia, L.; Gallo, J.; Bañobre-López, M.; Dražić, G.; Calhelha, R.C.; Ferreira, I.C.F.R.; Lima, R.; et al. Multifunctional graphene-based magnetic nanocarriers for combined hyperthermia and dual stimuli-responsive drug delivery. *Mater. Sci. Eng. C* **2018**, *93*, 206–217. [CrossRef] [PubMed]

17. Du, Y.; Liu, X.; Liang, Q.; Liang, X.J.; Tian, J. Optimization and Design of Magnetic Ferrite Nanoparticles with Uniform Tumor Distribution for Highly Sensitive MRI/MPI Performance and Improved Magnetic Hyperthermia Therapy. *Nano Lett.* **2019**, *19*, 3618–3626. [CrossRef] [PubMed]

18. Koh, I.; Josephson, L. Magnetic Nanoparticle Sensors. *Sensors* **2009**, *9*, 8130–8145. [CrossRef] [PubMed]

19. Jaufenthaler, A.; Schier, P.; Middelmann, T.; Liebl, M.; Wiekhorst, F.; Baumgarten, D. Quantitative 2D magnetorelaxometry imaging of magnetic nanoparticles using optically pumped magnetometers. *Sensors* **2020**, *20*, 753. [CrossRef] [PubMed]

20. Chieh, J.J.; Wei, W.C.; Liao, S.H.; Chen, H.H.; Lee, Y.F.; Lin, F.C.; Chiang, M.H.; Chiu, M.J.; Horng, H.E.; Yang, S.Y. Eight-channel AC magnetosusceptometer of magnetic nanoparticles for high-throughput and ultra-high-sensitivity immunoassay. *Sensors* **2018**, *18*, 1043. [CrossRef]

21. Périgo, E.A.; Hemery, G.; Sandre, O.; Ortega, D.; Garaio, E.; Plazaola, F.; Teran, F.J. Fundamentals and advances in magnetic hyperthermia. *Appl. Phys. Rev.* **2015**, *2*, 041302. [CrossRef]

22. Patil, R.M.; Thorat, N.D.; Shete, P.B.; Otari, S.V.; Tiwale, B.M.; Pawar, S.H. In vitro hyperthermia with improved colloidal stability and enhanced SAR of magnetic core/shell nanostructures. *Mater. Sci. Eng. C* **2016**, *59*, 702–709. [CrossRef]

23. Aquino, V.R.R.; Vinícius-Araújo, M.; Shrivastava, N.; Sousa, M.H.; Coaquira, J.A.H.; Bakuzis, A.F. Role of the Fraction of Blocked Nanoparticles on the Hyperthermia Efficiency of Mn-Based Ferrites at Clinically Relevant Conditions. *J. Phys. Chem. C* **2019**, *123*, 27725–27734. [CrossRef]

24. Maldonado-Camargo, L.; Torres-Díaz, I.; Chiu-Lam, A.; Hernández, M.; Rinaldi, C. Estimating the contribution of Brownian and Néel relaxation in a magnetic fluid through dynamic magnetic susceptibility measurements. *J. Magn. Magn. Mater.* **2016**, *412*, 223–233. [CrossRef]

25. Nemati, Z.; Alonso, J.; Rodrigo, I.; Das, R.; Garaio, E.; García, J.Á.; Orue, I.; Phan, M.H.; Srikanth, H. Improving the Heating Efficiency of Iron Oxide Nanoparticles by Tuning Their Shape and Size. *J. Phys. Chem. C* **2018**, *122*, 2367–2381. [CrossRef]

26. Bender, P.; Fock, J.; Frandsen, C.; Hansen, M.F.; Balceris, C.; Ludwig, F.; Posth, O.; Wetterskog, E.; Bogart, L.K.; Southern, P.; et al. Relating Magnetic Properties and High Hyperthermia Performance of Iron Oxide Nanoflowers. *J. Phys. Chem. C* **2018**, *122*, 3068–3077. [CrossRef]

27. Atkinson, W.; Brezovich, I.; Chakraborty, D.P. Usable Frequencies in Hyperthermia with Thermal Seeds. *IEEE Trans. Biomed. Eng.* **1984**, *31*, 70–75. [CrossRef] [PubMed]

28. Hergt, R.; Dutz, S. Magnetic particle hyperthermia—Biophysical limitations of a visionary tumour therapy. *J. Magn. Magn. Mater.* **2007**, *311*, 187–192. [CrossRef]

29. Shaterabadi, Z.; Nabiyouni, G.; Soleymani, M. Physics responsible for heating efficiency and self-controlled temperature rise of magnetic nanoparticles in magnetic hyperthermia therapy. *Prog. Biophys. Mol. Biol.* **2018**, *133*, 9–19. [CrossRef] [PubMed]

30. Reddy, L.H.; Arias, J.L.; Nicolas, J.; Couvreur, P. Magnetic Nanoparticles: Design and Characterization, Toxicity and Biocompatibility, Pharmaceutical and Biomedical Applications. *Chem. Rev.* **2012**, *112*, 5818–5878. [CrossRef]
31. Blanco-Andujar, C.; Walter, A.; Cotin, G.; Bordeianu, C.; Mertz, D.; Felder-Flesch, D.; Begin-Colin, S. Design of iron oxide-based nanoparticles for MRI and magnetic hyperthermia. *Nanomedicine* **2016**, *11*, 1889–1910. [CrossRef]
32. Seehra, M. *Magnetic Spinels: Synthesis, Properties and Applications*; IntechOpen: London, UK, 2017.
33. Smit, J.; Wijn, H.P.J. *Ferrites*; Cleaver-Hume Press Ltd: London, UK, 1959.
34. He, S.; Zhang, H.; Liu, Y.; Sun, F.; Yu, X.; Li, X.; Zhang, L.; Wang, L.; Mao, K.; Wang, G.; et al. Maximizing Specific Loss Power for Magnetic Hyperthermia by Hard–Soft Mixed Ferrites. *Small* **2018**, *14*, 1800135. [CrossRef]
35. Pilati, V.; Cabreira Gomes, R.; Gomide, G.; Coppola, P.; Silva, F.G.; Paula, F.L.O.; Perzynski, R.; Goya, G.F.; Aquino, R.; Depeyrot, J. Core/Shell Nanoparticles of Non-Stoichiometric Zn-Mn and Zn-Co Ferrites as Thermosensitive Heat Sources for Magnetic Fluid Hyperthermia. *J. Phys. Chem. C* **2018**, *122*, 3028–3038. [CrossRef]
36. Sharifi Dehsari, H.; Asadi, K. Impact of Stoichiometry and Size on the Magnetic Properties of Cobalt Ferrite Nanoparticles. *J. Phys. Chem. C* **2018**, *122*, 29106–29121. [CrossRef]
37. Demirci Dönmez, C.E.; Manna, P.K.; Nickel, R.; Aktürk, S.; Van Lierop, J. Comparative Heating Efficiency of Cobalt-, Manganese-, and Nickel-Ferrite Nanoparticles for a Hyperthermia Agent in Biomedicines. *ACS Appl. Mater. Interfaces* **2019**, *11*, 6858–6866. [CrossRef]
38. Albino, M.; Fantechi, E.; Innocenti, C.; López-Ortega, A.; Bonanni, V.; Campo, G.; Pineider, F.; Gurioli, M.; Arosio, P.; Orlando, T.; et al. Role of Zn2+ Substitution on the Magnetic, Hyperthermic, and Relaxometric Properties of Cobalt Ferrite Nanoparticles. *J. Phys. Chem. C* **2019**, *123*, 6148–6157. [CrossRef]
39. Amiri, S.; Shokrollahi, H. The role of cobalt ferrite magnetic nanoparticles in medical science. *Mater. Sci. Eng. C* **2013**, *33*, 1–8. [CrossRef] [PubMed]
40. Fantechi, E.; Innocenti, C.; Albino, M.; Lottini, E.; Sangregorio, C. Influence of cobalt doping on the hyperthermic efficiency of magnetite nanoparticles. *J. Magn. Magn. Mater.* **2015**, *380*, 365–371. [CrossRef]
41. Joshi, H.M.; Lin, Y.P.; Aslam, M.; Prasad, P.V.; Schultz-Sikma, E.A.; Edelman, R.; Meade, T.; Dravid, V.P. Effects of shape and size of cobalt ferrite nanostructures on their MRI contrast and thermal activation. *J. Phys. Chem. C* **2009**, *113*, 17761–17767. [CrossRef]
42. Mallick, A.; Mahapatra, A.S.; Mitra, A.; Greneche, J.M.; Ningthoujam, R.S.; Chakrabarti, P.K. Magnetic properties and bio-medical applications in hyperthermia of lithium zinc ferrite nanoparticles integrated with reduced graphene oxide. *J. Appl. Phys.* **2018**, *123*, 055103. [CrossRef]
43. *Lithium and Cell Physiology*; Bach, P.O.; Gallicchio, V.S. (Eds.) Springer: New York, NY, USA, 1990; ISBN 978-1-4612-7967-9.
44. Barrera, G.; Coisson, M.; Celegato, F.; Raghuvanshi, S.; Mazaleyrat, F.; Kane, S.N.; Tiberto, P. Cation distribution effect on static and dynamic magnetic properties of Co1-xZnxFe2O4ferrite powders. *J. Magn. Magn. Mater.* **2018**, *456*, 372–380. [CrossRef]
45. Chakrabarty, S.; Dutta, A.; Pal, M. Enhanced magnetic properties of doped cobalt ferrite nanoparticles by virtue of cation distribution. *J. Alloys Compd.* **2015**, *625*, 216–223. [CrossRef]
46. Mohamed, M.B.; Yehia, M. Cation distribution and magnetic properties of nanocrystalline gallium substituted cobalt ferrite. *J. Alloys Compd.* **2014**, *615*, 181–187. [CrossRef]
47. Thanh, N.K.; Loan, T.T.; Duong, N.P.; Anh, L.N.; Nguyet, D.T.T.; Nam, N.H.; Soontaranon, S.; Klysubun, W.; Hien, T.D. Cation Distribution Assisted Tuning of Magnetization in Nanosized Magnesium Ferrite. *Phys. Status Solidi Appl. Mater. Sci.* **2018**, *215*, 1700397. [CrossRef]
48. Tomitaka, A.; Hirukawa, A.; Yamada, T.; Morishita, S.; Takemura, Y. Biocompatibility of various ferrite nanoparticles evaluated by in vitro cytotoxicity assays using HeLa cells. *J. Magn. Magn. Mater.* **2009**, *321*, 1482–1484. [CrossRef]
49. Oliveira, A.B.B.; De Moraes, F.R.; Candido, N.M.; Sampaio, I.; Paula, A.S.; De Vasconcellos, A.; Silva, T.C.; Miller, A.H.; Rahal, P.; Nery, J.G.; et al. Metabolic Effects of Cobalt Ferrite Nanoparticles on Cervical Carcinoma Cells and Nontumorigenic Keratinocytes. *J. Proteome Res.* **2016**, *15*, 4337–4348. [CrossRef] [PubMed]

50. Giri, J.; Pradhan, P.; Somani, V.; Chelawat, H.; Chhatre, S.; Banerjee, R.; Bahadur, D. Synthesis and characterizations of water-based ferrofluids of substituted ferrites [$Fe_{1-x}B_xFe_2O_4$, B = Mn, Co (x = 0–1)] for biomedical applications. *J. Magn. Magn. Mater.* **2008**, *320*, 724–730. [CrossRef]

51. Amiri, M.; Salavati-Niasari, M.; Akbari, A. Magnetic nanocarriers: Evolution of spinel ferrites for medical applications. *Adv. Colloid Interface Sci.* **2019**, *265*, 29–44. [CrossRef] [PubMed]

52. Raghuvanshi, S.; Mazaleyrat, F.; Kane, S.N. Mg1-xZnxFe2O4nanoparticles: Interplay between cation distribution and magnetic properties. *AIP Adv.* **2018**, *8*, 047804. [CrossRef]

53. Patton, C.E.; Edmondson, C.A.; Liu, Y.H. Magnetic properties of lithium zinc ferrite. *J. Appl. Phys.* **1982**, *53*, 2431–2433. [CrossRef]

54. Lutterotti, L.; Scardi, P. Simultaneous structure and size-strain refinement by the Rietveld method. *J. Appl. Crystallogr.* **1990**, *23*, 246–252. [CrossRef]

55. Weil, L.; Bertaut, F.; Bochirol, L.; Weil, L.; Bertaut, F.; Propri, L.B. Propriétés magnétiques et structure de la phase quadratique du ferrite de cuivre. *J.Phys. Radium* **1950**, *11*, 208–212. [CrossRef]

56. Tanna, A.R.; Joshi, H.H. Computer Aided X-Ray Diffraction Intensity Analysis for Spinels: Hands-On Computing Experience. *Int. J. Phys. Math. Sci.* **2013**, *7*, 334–340.

57. Wolska, E.; Riedel, E.; Wolski, W. The Evidence of $Cd^{2+}{}_xFe_{1-x}{}^{3+}[Ni_{1-x}{}^{2+}Fe_{1+x}{}^{3+}]O_4$ Cation Distribution Based on X-Ray and Mössbauer Data. *Phys. Status Solidi* **1992**, *132*, K51–K56. [CrossRef]

58. Červinka, L.; Šimša, Z. Distribution of copper ions in some copper-manganese ferrites. *Czechoslov. J. Phys.* **1970**, *20*, 470–474. [CrossRef]

59. Cullity, B.D.; Graham, C.D. *Introduction to Magnetic Materials*; John Wiley & Sons: Hoboken, NJ, USA, 2009; ISBN 9780471477419.

60. Coïsson, M.; Barrera, G.; Celegato, F.; Martino, L.; Kane, S.N.; Raghuvanshi, S.; Vinai, F.; Tiberto, P. Hysteresis losses and specific absorption rate measurements in magnetic nanoparticles for hyperthermia applications. *Biochim. Biophys. Acta Gen. Subj.* **2017**, *1861*, 1545–1558. [CrossRef] [PubMed]

61. Kurlyandskaya, G.V.; Litvinova, L.S.; Safronov, A.P.; Schupletsova, V.V.; Tyukova, I.S.; Khaziakhmatova, O.G.; Slepchenko, G.B.; Yurova, K.A.; Cherempey, E.G.; Kulesh, N.A.; et al. Water-Based suspensions of iron oxide nanoparticles with electrostatic or steric stabilization by chitosan: Fabrication, characterization and biocompatibility. *Sensors* **2017**, *17*, 2605. [CrossRef] [PubMed]

62. Ranjith Kumar, E.; Jayaprakash, R.; Kumar, S. The role of annealing temperature and bio template (egg white) on the structural, morphological and magnetic properties of manganese substituted MFe_2O_4(M = Zn, Cu, Ni, Co) nanoparticles. *J. Magn. Magn. Mater.* **2014**, *351*, 70–75. [CrossRef]

63. Jadhav, S.A. Magnetic properties of Zn-substituted Li-Cu ferrites. *J. Magn. Magn. Mater.* **2001**, *224*, 167–172. [CrossRef]

64. Patil, R.S.; Kakatkar, S.V.; Patil, S.A.; Sankpal, A.M.; Sawant, S.R. X-Ray and bulk magnetic studies on $Li_{0.5}Zn_xTi_xFe_{2.5-2x}O_4$. *Mater. Chem. Phys.* **1991**, *28*, 355–365. [CrossRef]

65. Gore, S.K.; Jadhav, S.S.; Jadhav, V.V.; Patange, S.M.; Naushad, M.; Mane, R.S.; Kim, K.H. The structural and magnetic properties of dual phase cobalt ferrite. *Sci. Rep.* **2017**, *7*, 1–9. [CrossRef]

66. Denton, A.R.; Ashcroft, N.W. Vegard's law. *Phys. Rev. A* **1991**, *43*, 3161–3164. [CrossRef]

67. Satalkar, M.; Kane, S.N.; Ghosh, A.; Ghodke, N.; Barrera, G.; Celegato, F.; Coisson, M.; Tiberto, P.; Vinai, F. Synthesis and soft magnetic properties of $Zn_{0.8-x}Ni_xMg_{0.1}Cu_{0.1}Fe_2O_4$ (x = 0.0–0.8) ferrites prepared by sol-gel auto-combustion method. *J. Alloys Compd.* **2015**, *615*, S313–S316. [CrossRef]

68. Rathod, V.; Anupama, A.V.; Jali, V.M.; Hiremath, V.A.; Sahoo, B. Combustion synthesis, structure and magnetic properties of Li-Zn ferrite ceramic powders. *Ceram. Int.* **2017**, *43*, 14431–14440. [CrossRef]

69. Patterson, A.L. The Scherrer Formula for X-Ray Particle Size Determination. *Phys. Rev.* **1939**, *56*, 978. [CrossRef]

70. Randhawa, B.S.; Dosanjh, H.S.; Kumar, N. Synthesis of lithium ferrite by precursor and combustion methods: A comparative study. *J. Radioanal. Nucl. Chem.* **2007**, *274*, 581–591. [CrossRef]

71. Murugesan, C.; Perumal, M.; Chandrasekaran, G. Structural, dielectric and magnetic properties of cobalt ferrite prepared using auto combustion and ceramic route. *Phys. B Condens. Matter* **2014**, *448*, 53–56. [CrossRef]

72. Rathod, V.; Anupama, A.V.; Kumar, R.V.; Jali, V.M.; Sahoo, B. Correlated vibrations of the tetrahedral and octahedral complexes and splitting of the absorption bands in FTIR spectra of Li-Zn ferrites. *Vib. Spectrosc.* **2017**, *92*, 267–272. [CrossRef]

73. Mathew, D.S.; Juang, R.S. An overview of the structure and magnetism of spinel ferrite nanoparticles and their synthesis in microemulsions. *Chem. Eng. J.* **2007**, *129*, 51–65. [CrossRef]

74. Coey, J.M.D. *Magnetism and Magnetic Materials*; Cambridge University Press: Cambridge, UK, 2009; ISBN 9780521816144.

75. Tachiki, M. Origin of the Magnetic Anisotropy Energy of Cobalt Ferrite. *Prog. Theor. Phys.* **1960**, *23*, 1055–1072. [CrossRef]

76. Fairweather, A.; Roberts, F.F.; Welch, A.J.E. Ferrites. *Reports Prog. Phys.* **1952**, *15*, 306. [CrossRef]

77. Murthy, N.S.S.; Natera, M.G.; Youssef, S.I.; Begum, R.J.; Srivastava, C.M. Yafet-kittel angles in zinc-nickel ferrites. *Phys. Rev.* **1969**, *181*, 969–977. [CrossRef]

78. El-Sayed, H.M.; Ali, I.A.; Azzam, A.; Sattara, A.A. Influence of the magnetic dead layer thickness of Mg-Zn ferrites nanoparticle on their magnetic properties. *J. Magn. Magn. Mater.* **2017**, *424*, 226–232. [CrossRef]

79. Sciancalepore, C.; Gualtieri, A.F.; Scardi, P.; Flor, A.; Allia, P.; Tiberto, P.; Barrera, G.; Messori, M.; Bondioli, F. Structural characterization and functional correlation of Fe_3O_4 nanocrystals obtained using 2-ethyl-1,3-hexanediol as innovative reactive solvent in non-hydrolytic sol-gel synthesis. *Mater. Chem. Phys.* **2018**, *207*, 337–349. [CrossRef]

80. Reilly, J.P. Principles of Nerve and Heart Excitation by Time-varying Magnetic Fields. *Ann. N. Y. Acad. Sci.* **1992**, *649*, 96–117. [CrossRef] [PubMed]

81. Oleson, J.R.; Cetas, T.C.; Corry, P.M. Hyperthermia by Magnetic Induction: Experimental and Theoretical Results for Coaxial Coil Pairs. *Radiat. Res.* **1983**, *95*, 175. [CrossRef] [PubMed]

82. Beola, L.; Gutiérrez, L.; Grazú, V.; Asín, L. A Roadmap to the Standardization of In Vivo Magnetic Hyperthermia. *Nanomater. Magn. Opt. Hyperth. Appl.* **2019**, 317–337.

83. Garaio, E.; Sandre, O.; Collantes, J.M.; Garcia, J.A.; Mornet, S.; Plazaola, F. Specific absorption rate dependence on temperature in magnetic field hyperthermia measured by dynamic hysteresis losses (ac magnetometry). *Nanotechnology* **2015**, *26*, 015704. [CrossRef]

84. Garaio, E.; Collantes, J.M.; Garcia, J.A.; Plazaola, F.; Mornet, S.; Couillaud, F.; Sandre, O. A wide-frequency range AC magnetometer to measure the specific absorption rate in nanoparticles for magnetic hyperthermia. *J. Magn. Magn. Mater.* **2014**, *368*, 432–437. [CrossRef]

85. Guibert, C.; Fresnais, J.; Peyre, V.; Dupuis, V. Magnetic fluid hyperthermia probed by both calorimetric and dynamic hysteresis measurements. *J. Magn. Magn. Mater.* **2017**, *421*, 384–392. [CrossRef]

86. O'Connell, J. Heating water: Rate correction due to Newtonian cooling. *Phys. Teach.* **1999**, *37*, 551–552. [CrossRef]

87. Dutz, S.; Hergt, R. Magnetic particle hyperthermia—A promising tumour therapy? *Nanotechnology* **2014**, *25*, 452001. [CrossRef]

88. Ota, S.; Takemura, Y. Characterization of Néel and Brownian Relaxations Isolated from Complex Dynamics Influenced by Dipole Interactions in Magnetic Nanoparticles. *J. Phys. Chem. C* **2019**, *123*, 28859–28866. [CrossRef]

89. Soukup, D.; Moise, S.; Céspedes, E.; Dobson, J.; Telling, N.D. *In Situ* Measurement of Magnetization Relaxation of Internalized Nanoparticles in Live Cells. *ACS Nano* **2015**, *9*, 231–240. [CrossRef]

90. Di Corato, R.; Espinosa, A.; Lartigue, L.; Tharaud, M.; Chat, S.; Pellegrino, T.; Ménager, C.; Gazeau, F.; Wilhelm, C. Magnetic hyperthermia efficiency in the cellular environment fordifferent nanoparticle designs. *Biomaterials* **2014**, *35*, 6400–6411. [CrossRef] [PubMed]

91. Kalambur, V.S.; Han, B.; Hammer, B.E.; Shield, T.W.; Bischof, J.C. In vitro characterization of movement, heating and visualization of magnetic nanoparticles for biomedical applications. *Nanotechnology* **2005**, *16*, 1221–1233. [CrossRef]

92. Coral, D.F.; Mendoza Zélis, P.; Marciello, M.; Morales, M.D.P.; Craievich, A.; Sánchez, F.H.; Fernández Van Raap, M.B. Effect of Nanoclustering and Dipolar Interactions in Heat Generation for Magnetic Hyperthermia. *Langmuir* **2016**, *32*, 1201–1213. [CrossRef] [PubMed]

93. Serantes, D.; Baldomir, D.; Martinez-Boubeta, C.; Simeonidis, K.; Angelakeris, M.; Natividad, E.; Castro, M.; Mediano, A.; Chen, D.X.; Sanchez, A.; et al. Influence of dipolar interactions on hyperthermia properties of ferromagnetic particles. *J. Appl. Phys.* **2010**, *108*, 073918. [CrossRef]

94. Branquinho, L.C.; Carrião, M.S.; Costa, A.S.; Zufelato, N.; Sousa, M.H.; Miotto, R.; Ivkov, R.; Bakuzis, A.F. Effect of magnetic dipolar interactions on nanoparticle heating efficiency: Implications for cancer hyperthermia. *Sci. Rep.* **2013**, *3*, 20–22. [CrossRef]

95. Dalal, M.; Ningthoujam, R.S.; Chakrabarti, P.K. Structural, magnetic, microwave and ac induction heating study of Li0.35Zn0.30Co0.05Fe2.3O4 integrated in multi-walled carbon nanotube matrix. *AIP Conf. Proc.* **2018**, *1942*, 3–7.

96. Wang, L.; Yan, Y.; Wang, M.; Yang, H.; Zhou, Z.; Peng, C.; Yang, S. An integrated nanoplatform for theranostics via multifunctional core-shell ferrite nanocubes. *J. Mater. Chem. B* **2016**, *4*, 1908–1914. [CrossRef]

97. Thorat, N.D.; Bohara, R.A.; Yadav, H.M.; Tofail, S.A.M. Multi-modal MR imaging and magnetic hyperthermia study of Gd doped Fe3O4 nanoparticles for integrative cancer therapy. *RSC Adv.* **2016**, *6*, 94967–94975. [CrossRef]

98. Kim, D.H.; Lee, S.H.; Kim, K.N.; Kim, K.M.; Shim, I.B.; Lee, Y.K. Temperature change of various ferrite particles with alternating magnetic field for hyperthermic application. *J. Magn. Magn. Mater.* **2005**, *293*, 320–327. [CrossRef]

99. Veverka, P.; Pollert, E.; Závěta, K.; Vasseur, S.; Duguet, E. Sr-hexaferrite/maghemite composite nanoparticles—Possible new mediators for magnetic hyperthermia. *Nanotechnology* **2008**, *19*, 215705. [CrossRef]

100. Lee, J.H.; Jang, J.T.; Choi, J.S.; Moon, S.H.; Noh, S.H.; Kim, J.W.; Kim, J.G.; Kim, I.S.; Park, K.I.; Cheon, J. Exchange-coupled magnetic nanoparticles for efficient heat induction. *Nat. Nanotechnol.* **2011**, *6*, 418–422. [CrossRef] [PubMed]

101. Hergt, R.; Hiergeist, R.; Zeisberger, M.; Schüler, D.; Heyen, U.; Hilger, I.; Kaiser, W.A. Magnetic properties of bacterial magnetosomes as potential diagnostic and therapeutic tools. *J. Magn. Magn. Mater.* **2005**, *293*, 80–86. [CrossRef]

Article

Real Time Monitoring of Calcium Oxalate Precipitation Reaction by Using Corrosion Resistant Magnetoelastic Resonance Sensors

Beatriz Sisniega [1,*]**, Ariane Sagasti Sedano** [1]**, Jon Gutiérrez** [1,2] **and Alfredo García-Arribas** [1,2]

[1] Departament de Electricidad y Electrónica, Universidad del País Vasco/Euskal Herriko Unibertsitatea (UPV/EHU), Barrio Sarriena s/n, 48940 Leioa, Spain; ariane.sagastis@ehu.eus (A.S.S.); jon.gutierrez@ehu.eus (J.G.); alfredo.garcia@ehu.es (A.G.-A.)

[2] BC Materials, Basque Center for Materials, Applications and Nanostructures, UPV/EHU Science Park, 48940 Leioa, Spain

*   Correspondence: beatriz.sisniega@ehu.eus

Received: 31 March 2020; Accepted: 11 May 2020; Published: 14 May 2020

Abstract: The magnetoelastic resonance is used to monitor the precipitation reaction of calcium oxalate (CaC_2O_4) crystals in real-time, by measuring the shift of the resonance frequency caused by the mass increase on the resonator. With respect to previous work on the same matter, the novelty lies in the adoption of an amorphous ferromagnetic alloy, of composition $Fe_{73}Cr_5Si_{10}B_{12}$, as resonator, that replaces the commercial Metglas® 2826 alloy (composition $Fe_{40}Ni_{38}Mo_4B_{18}$). The enhanced corrosion resistance of this material allows it to be used in biological environments without any pre-treatment of its surface. Additionally, the measurement method, which has been specifically adapted to this application, allows quick registration of the whole resonance curve as a function of the excitation frequency, and thus enhances the resolution and decreases the detection noise. The frequency shift is calibrated by the static deposition of well-known masses of CaC_2O_4. The resonator dimensions have been selected to improve sensitivity. A 20 mm long, 2 mm wide and 25 µm thick magnetoelastic resonator has been used to monitor the precipitation reaction of calcium oxalate in a 500 s time interval. The results of the detected precipitated mass when oxalic acid and calcium chloride are mixed in different concentrations (30 mM, 50 mM and 100 mM) are presented as a function of time. The results show that the sensor is capable of monitoring the precipitation reaction. The mass sensitivity obtained, and the corrosion resistance of the material, suggest that this material can perform excellently in monitoring this type of reaction.

Keywords: magnetoelasticity; precipitation; mass measurement; chemical sensor

---

1. Introduction

Magnetoelastic resonance sensors are typically made of amorphous ferromagnetic ribbons, with good values of spontaneous magnetization and saturation magnetostriction, and low magnetocrystalline anisotropy [1–3]. In this type of material the mechanical and magnetic properties are intimately coupled by magnetostriction, and so an acoustic wave can be excited within the material by the application of an alternating magnetic field. Reciprocally, mechanical disturbances, such as oscillations, can be magnetically detected by the voltage induced in a coil in the proximity of the sample. The magnetoelastic material, which is usually fabricated in the form of a thin ribbon, can enter in resonance at certain frequencies of excitation, compatible with the dimensions and elastic properties of the material. The resonance is extremely sensitive to different external parameters, which can be used to design different types of sensors [4,5]. In particular, differences in mass loading (Δm) cause a variation of the resonance frequency ($\Delta f = f_r - f_0$) of a bare magnetoelastic ribbon of mass m_0, determined by the expression [5]:

$$\Delta f = -\frac{1}{2} f_0 \left(\frac{\Delta m}{m_0} \right) \tag{1}$$

The sensitivity to these external parameters, together with the ability to query and detect remotely, make these devices especially interesting for sensing biological and chemical agents.

Previous works by Boropoulos and co-workers [6] have used magnetoelastic sensors, based on the commercial material Metglas® 2826, for monitoring the kinetics of different precipitation reactions, such as the precipitation of calcium oxalate (CaC_2O_4) crystals, one of the most common minerals that form calcifications in the urinary tract (so-called kidney or bladder stones). In humans, there are essential inorganic salts for diverse metabolic activities, like sodium chloride ($NaCl$), calcium chloride ($CaCl_2$), magnesium chloride ($MgCl_2$), sodium bicarbonate ($NaHCO_3$), potassium chloride (KCl), sodium sulfate (Na_2SO_4), calcium carbonate ($CaCO_3$), and calcium phosphate ($Ca_3(PO_4)_2$). These inorganic salts dissociate in solution into ions (or electrolytes). If some of these ions are not properly absorbed within the body, they will tend to crystallize in small grains or stones.

Concerning kidney stones, their formation is the result of a complex physicochemical process that leads to crystallization. The stones, also known as renal calculi, are mostly calcium based [7]: calcium oxalate accounts for approximately 60% to 70%, struvite (magnesium ammonium phosphate) for 10% to 20%, uric acid for 5% to 10%, and in low quantities also cystine (<1%) and calcium phosphate (<5%). Some conditions, such as hypercalciuria or hyperoxaluria, can contribute to increasing the risk of suffering this pathology. Hypercalciuria is the result of an increase in the quantity of filtered calcium and a decrease of its re-absorption in the kidney. It is thought to contribute to kidney stone-formation by creating a urine supersaturated with respect to calcium. Supersaturation arises from a concentration above a material's solubility in water, and leads to the formation of crystals. In a urine sample collected over 24 h from an average adult, a quantity of 100–250 mg of calcium is expected. In conditions of hypercalciuria the urine calcium excretion is greater than 275–300 mg/day in men, or 250 mg/day in women. Hypercalciuria can also be defined as a daily urinary excretion of more than 4 mg calcium/kg body weight [8,9]. In the adult population, 5% of men have hypercalciuria and, of them, about 10% will form a kidney stone. Urine has a remarkable ability to inhibit calcium crystallization, which thus prevents most of the population from continuously forming such stones. Hyperoxaluria, on the other hand, is a metabolic disorder of increased excretion of oxalate in urine, and it is also related to the formation of stones in the urinary tract [10]. The normal values of oxalate excretion in urine are under 40 mg/day [11]. Urine calcium oxalate concentration values up to 100 µmol/L are reported in patients suffering from primary hyperoxaluria [12].

The use of magnetoelastic sensors to remotely monitor these types of reactions (deposition of crystallites) can provide fundamental information about the precipitation process in biological fluids. It can increase our understanding of these complex processes of bio-mineralization since this technique allows, for example, to study precipitation systems under the influence of different factors (such as pH, or concentration and chemical composition of the urine), to enhance our knowledge about which factors or substances favour or inhibit the formation of crystals in the urinary tract [13,14]. Information about the precipitate mass in real-time can complement the information (usually obtained by monitoring changes in pH or concentration) of studies in artificial systems that mimic the human physiological conditions, like human urine [14].

In the mentioned work [6], the formation of the insoluble salt crystals was tracked by the changes in the resonant frequency of the magnetoelastic sensor, which decreases as the precipitate is deposited on its surface, as a direct consequence of the increase of the total mass of the resonant strip. In spite of the excellent results published in that work, there are two main aspects that, in our opinion, can be improved. First, the Metglas® 2826 alloy used as magnetoelastic resonator had to be submitted to a previous conditioning process to protect it from corrosion occurring in the biological medium. Second, the resonance changes due to precipitation were followed by measuring the voltage induced at a fixed frequency (117 kHz, above the resonant frequency), in order to be able to follow the process in real-time, a procedure that reduces the accuracy and resolution of the measurement.

In the present work, we have reproduced the process of monitoring the precipitation reaction of calcium oxalate crystals, but in this case, we have used an amorphous ferromagnetic alloy of composition $Fe_{73}Cr_5Si_{10}B_{12}$ as resonator. This alloy does not need any special pre-treatment, since·it has been already demonstrated to exhibit excellent resistance to corrosion [15,16]. Additionally, the measurement process has been optimized, so we can measure the complete resonance curve fast enough to directly track the evolution of the resonance frequency along the precipitation process. The change in the resonance frequency during the precipitation process has been directly correlated to the amount of precipitated material on the resonator through a calibration with known deposited masses. We have analyzed carefully the validity of this calibration, and concluded that, even though the calibration is performed with the resonator vibrating in air, it provides a valid estimation of the mass of precipitated materials in the liquid phase.

2. Materials and Methods

2.1. Magnetic and Magnetoelastic Materials Characterization

The magnetoelastic ribbon of composition $Fe_{73}Cr_5Si_{10}B_{12}$ used in this work was kindly provided by Vacuumschmelze GmbH & Co., KG, Germany. This composition contains a small amount of chromium (5% atomic) that allows the formation of a passivation layer on the material, thus favoring its corrosion-resistant behavior. We selected the geometry of the strips, cut from a larger ribbon, to be 20 mm × 2 mm × 25 μm, with a length to width ratio ($R = L/w = 10$), high enough to ensure a good magnetoelastic coupling [17] and enough surface for the deposition of precipitate.

Magnetic and corrosion resistance characterization appear extensively explained in [16]. Magnetoelastic measurements, for the real-time monitoring of the resonance frequency, were carried out by using a homemade experimental set-up [18]. Briefly, the system used to register the resonance–antiresonance curve consists of three coaxial solenoids: one to apply the constant bias field (H); a second one to produce the alternating field to magnetostrictively excite the sample; and the third one, consisting of a compensated pick-up coil, to detect the induced magnetization oscillations, from which the frequencies of the corresponding magnetoelastic resonance (f_r) and anti-resonance (f_a) are determined. We optimized the data acquisition procedure to be able to register the whole resonance–antiresonance curve quick enough to follow the precipitation process. A spectrum analyzer (HP 3589A) working in swept mode was used to produce the excitation and to receive the signal induced in the pick-up coil. The range of the frequency sweep was kept as small as possible. It was determined prior to the experiment, based on the expected change of the resonance frequency. The speed of the sweep was set as fast as possible, while maintaining a compromise with the quality of the registered curve. The goal was to allow discriminating differences in the resonance frequency of 100 Hz. After recording the resonance–antiresonance, the frequency of the maximum and its amplitude were measured using the built-in analysis procedures of the analyzer, and transmitted to a control computer. The typical time to record a whole magnetoelastic resonance curve was about 5 s.

The measured resonant frequency (f_r) varies with the bias field H, since it is directly related to Young's modulus as $f_r = \left(\sqrt{E(H)/\rho}\right)/2L$ [19], where L and ρ are the length and density of the ribbon shaped material. The field-dependence of the elastic modulus is known as ΔE effect and quantified as $\Delta E(\%) = (1 - E(H)/E_S) \times 100$, E_S being the Young's modulus measured at magnetic saturation. Other important information that can be determined from these measurements are the magnetoelastic coupling coefficient ($k^2 = (\pi^2/8)(1 - (f_r/f_a)^2)$) [20] and the quality factor of the resonance ($Q = f_r/\Delta f$), all quantities being functions of the applied external magnetic field.

Table 1 shows the measured magnetic, magnetoelastic and corrosion resistance properties of the $Fe_{73}Cr_5Si_{10}B_{12}$ ribbon used in our calcium oxalate detection experiments.

Table 1. Magnetic and magnetoelastic parameters of the $Fe_{73}Cr_5Si_{10}B_{12}$ sample. The ones for Metglas® 2826 alloy are also shown for comparison. Data taken from ref. [16].

Composition	$\mu_0 M_s$ (T)	λ_s (ppm)	ΔE (%)	k	E_{corr} (mV)	Corrosion Rate (μm/year)
$Fe_{73}Cr_5Si_{10}B_{12}$	1.12	14	17	0.41	47	0.035
$Fe_{40}Ni_{38}Mo_4B_{18}$ Metglas® 2826 *	0.88	11	2.5	0.16	−427	23.4

* Commercially available magnetoelastic ribbon [21,22].

2.2. Sensor Calibration

Prior to the calibration of the sensor, we studied the influence that the medium, in which the magnetoelastic strip is immersed (air or distilled water), has on the observed magnetoelastic resonance. Figure 1 shows both the change in one single magnetoelastic resonance curve, and over the whole measured $f_r(H)$ dependence, from zero to saturation applied H magnetic fields.

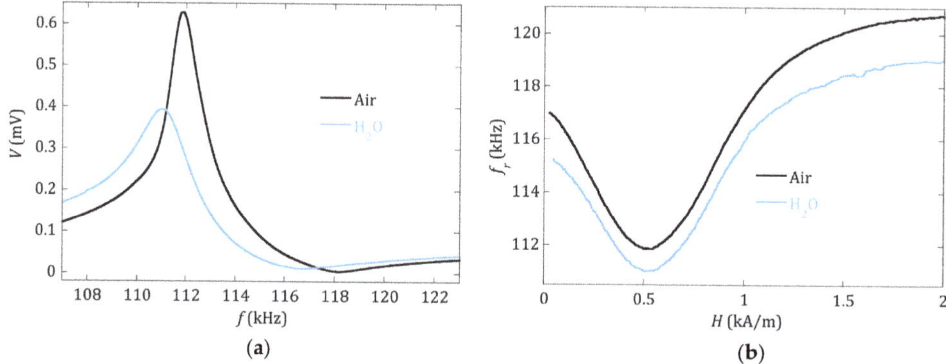

Figure 1. (a) Magnetoelastic resonance curves measured at H = 517 A/m, and (b) dependence of the resonant frequency (f_r) with the applied magnetic field, for the $Fe_{73}Cr_5Si_{10}B_{12}$ strip measured in air and when it is immersed in distilled water.

As expected, and due to the differences in viscosity and density of air and distilled water, the effect of the immersion in water is to increase the damping in a uniform way over the whole strip. This means that, at the same applied bias field H, the magnetoelastic resonance curve widens (the quality factor Q value decreases) and the resonance frequency decreases (see Figure 1a). The $f_r(H)$ curves displayed in Figure 1b show that the decrease of the resonance frequency is almost the same at any applied H magnetic field, at least in the vicinity of the minimum, which is where the measurements are made during the precipitation experiment.

The calibration of the sensor sensitivity to added mass was performed by depositing in successive steps a known precipitated mass on the sensor and measuring the corresponding change in its magnetoelastic resonance frequency. The deposition process for the calibration is the same used for the calcium oxalate detection measurement conditions: the sensor is immersed in a small vial with a mixture of the precipitation solution with different concentrations, and once the precipitated crystallites are deposited onto it, it is taken out of the vial carefully and dried. Afterwards, the sensor is weighed on a precision balance (0.1 μg resolution), and its magnetoelastic resonant frequency is measured at a bias field corresponding to the minimum of the $f_r(H)$ curve. This minimum corresponds to a bias field equal to the anisotropy field (H_k) of the resonator, which in our case is H_k = 517 A/m. At this anisotropy field value, the sensitivity of the magnetoelastic resonance frequency to the amount of added mass is maximum [23].

For this calibration, Equation (1) is just an approximation of the more general expression [5]:

$$\frac{f_r}{f_0} = \left(1 + \frac{\Delta m}{m_0}\right)^{-1/2} \tag{2}$$

The mass and resonance frequency of the bare magnetoelastic sensor strip are m_0 = 7.6063 mg and f_0 = 115.38 kHz, respectively. The mass changes suffered by the sensor during the calcium oxalate precipitation process are greater than 5% of that initial bare weight. Therefore, a second order expansion of Equation (2) has been used to obtain an appropriate fit for the calibration curve [24]:

$$\Delta f = f_r - f_0 = -\frac{f_0}{2m_0}\Delta m + \frac{3f_0}{8m_0^2}(\Delta m)^2 = a_1 \Delta m + a_2 (\Delta m)^2 \tag{3}$$

Following this procedure, the mass calibration allows a quantitative knowledge of the mass of precipitate deposited on the sensor during the precipitation process, and therefore, a real-time monitoring of the reaction kinetics.

The obtained calibration curve can be seen in Figure 2, and the experimentally obtained calibration constants are a_1 = −9.8 ± 0.4 kHz/mg and a_2 = 1.1 ± 0.3 kHz/mg^2. It is to be noted that, for the calibration, the measurement of the frequency shift caused by the precipitated mass on the resonator is made in air, whereas in the real-time experiments described in Section 2.3, the resonance takes place inside the water-filled vial. As evidenced in Figure 1b, the resonance frequency in water is systematically lower than the one in air. Grimes et al. [4] gave already an expression for the observed decrease in the magnetoelastic resonance frequency when the vibrating sensor is immersed in a viscous liquid:

$$\Delta f = -\frac{\sqrt{\pi \eta \rho_l}}{2\pi d \rho_s}(f_0)^{1/2} \tag{4}$$

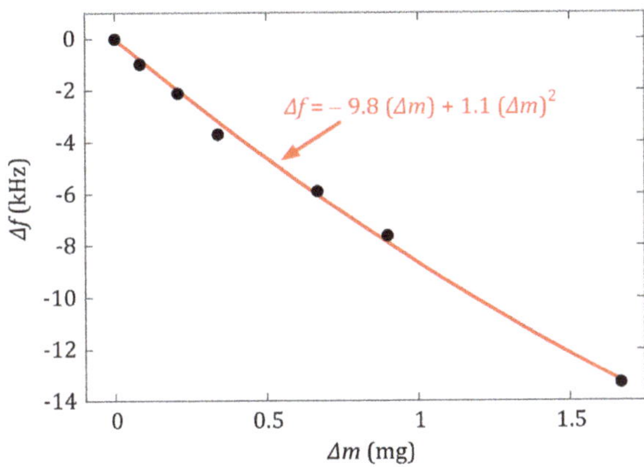

Figure 2. Calibration curve obtained from the changes in the resonance frequency of the $Fe_{73}Cr_5Si_{10}B_{12}$ resonator (measured in air), caused by different calcium oxalate mass depositions on its surface. Black dots represent the measured calibration points. The solid red line represents a fit to the second order expression of Equation (3), with coefficients a_1 = −9.8 kHz/mg and a_2 = 1.1 kHz/mg^2.

In our case, η = 0.89 × 10^{-3} Pa·s and ρ_l = 1000 kg/m^3 are the viscosity and density of water, and ρ_s = 7200 kg/m^3 and d = 25 μm are the density and thickness of the resonant strip, respectively. If we apply the frequency decrease established by Equation (4) to the resonance frequency values in the calibration curve, we obtain the same values, within the error, of the fitting parameters a_1 and a_2. Therefore, we conclude that the calibration curve measured in air is a good approximation to determine the amount of calcium oxalate mass deposited in the real-time experiments, when the resonant strip is immersed inside the solutions.

Finally, the theoretical values of the mass calibration constants appearing in Equation (3) can be calculated to be $a_1 = -\frac{f_0}{2m_0} = -7.6\,\text{kHz/mg}$ and $a_2 = \frac{3f_0}{8m_0^2} = 0.8\,\text{kHz/mg}^2$. It can be observed that the experimental calibration constants are both 22%–27% higher than these expected values. This kind of discrepancy is observed in other works [24,25]. The origin of the deviation is not clear, but can be related to two aspects: first, an original oversimplification in the derivation of Equation (2), that identifies the increase of mass as a change in the resonator density [5]. Second, the fact that the apparent resonance frequency (maximum of the resonance curve) is shifted to lower frequencies when the damping in the resonance is increased. Equation (2) does not consider this effect; however, experimental evidence shows that mass loading increases the damping of the resonance (as clearly shown in Figure 3). Comparing with the sensor sensitivity reported by Bouropoulos et al. [6], $-1.38\,\text{kHz/mg}$, the main mass calibration constant (a_1) obtained for the $Fe_{73}Cr_5Si_{10}B_{12}$ magnetoelastic resonator in this work is about seven times more sensitive. The main reason that accounts for this fact is the better magnetoelastic coupling coefficient (k) of the $Fe_{73}Cr_5Si_{10}B_{12}$ alloy than the Metglas 2826 one, as can be seen in the characteristics given in Table 1. This is directly related to the length-to-width ratio ($R = L/w$) chosen for the resonator used in our experiments ($R = 10$, instead of $R \sim 3$ of the previous work [6]). According to [17], the magnetoelastic coupling coefficient reaches its optimal value at $R \geq 12$, so in our case the aspect ratio is high enough to ensure good magnetoelastic properties and quality factor (Q) of the sensor.

2.3. Calcium Oxalate Precipitation

The monitoring of the precipitation reaction was carried out by placing the magnetoelastic strip inside a vial and mixing equal parts of solutions (0.6 mL each) with the same concentration of oxalic acid and calcium chloride, leading to the formation of the insoluble crystals:

$$CaCl_2(aq) + H_2C_2O_4(aq) \rightarrow CaC_2O_4(s) + 2HCl(aq)$$

Solutions of different concentration (30 mM, 50 mM and 100 mM) were used in order to observe its effect on the rate of reaction, and to assess the detection capability of the technique. Precipitation of the calcium oxalate salt will occur immediately after the mixing of the solutions if, with respect to the solid phases, the resulting solution concentration turns out to be supersaturated (SI > 0). SI denotes the saturation index of a sparingly soluble salt in an aqueous medium [26,27]. On the contrary, precipitation will not occur for undersaturated (SI < 0) concentrations.

Prior to the time monitoring of the calcium oxalate salt precipitation process, a control curve of the sensing strip within the vial with only distilled water was also performed.

3. Results and Discussion

The effect of the different concentrations of the prepared precipitation solutions can be directly inferred from Figures 3 and 4. Figure 3 shows, in the same time window from 10 to 500 s, how quickly both the magnetoelastic resonance frequency (in kHz) and the amplitude of the detected signal (in mV) decreased as the concentration of the constituents within the solutions increased. The rate at which this decrease occurred was clearly higher for the 100 mM than for the 30 mM concentration solution.

Figure 4 shows the temporal evolution of the amplitude of the detected signal, as the precipitation reaction occurs. The linear fit of the initial slope of each measured curve indicates how fast the deposition process took place. In our study, and for the 30, 50 and 100 mM concentration cases, we got a signal decrease ratio of $-0.95\,\mu\text{V/s}$, $-1.76\,\mu\text{V/s}$ and $-4.5\,\mu\text{V/s}$, respectively. These ratios reflect how the precipitation process of the calcium oxalate took place, indicating that it was clearly different between the 30 mM concentration solutions and the 50 and 100 mM concentration ones. For these last two cases, there was a fast decrease in the measured signal, motivated by the high supersaturated character of both solutions, and subsequent spontaneous and quick formation of the salt. The higher the slope of the initial part of the curve, the faster the precipitation process took place.

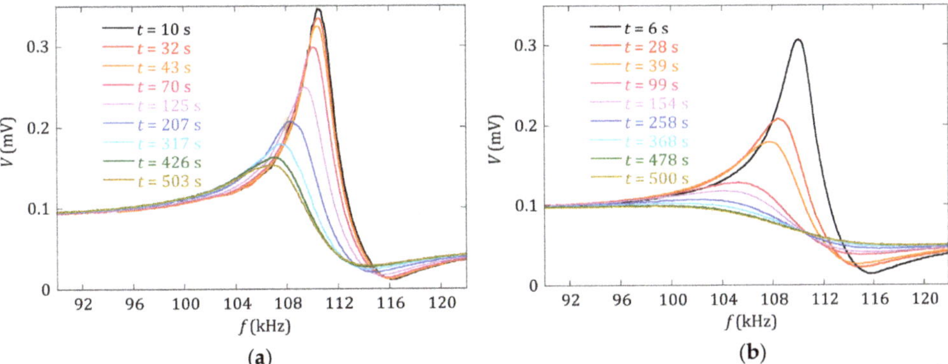

Figure 3. Measured magnetoelastic resonance curves of the sensor at different times during the precipitation process for solutions of oxalic acid and calcium chloride with concentrations of (**a**) 30 mM and (**b**) 100 mM.

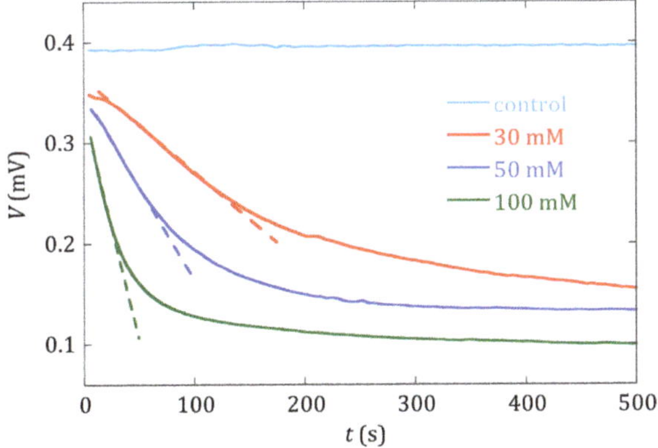

Figure 4. Temporal evolution of the amplitude (mV) of the measured signal (resonance amplitude), as the precipitation reaction progressed and precipitate crystals were deposited on the sensor surface. Curves are shown for different reactant concentrations (30 mM, 50 mM and 100 mM), and a control test (sensor in a vial with distilled water). Linear fits of the initial slope of each voltage curve are shown in dashed lines.

As remarked previously by other authors, when the precipitation of the salt happens, crystal growth is prevalent with receding nucleation [28]. In a final step, the decrease of the supersaturation character of the solutions leads to a plateau-like regime, with ratios −17.1 nV/s and −29 nV/s for the 50 and 100 mM concentration cases (300 s < time window < 500 s), indicating that the calcium oxalate precipitation process is practically finished.

On the contrary, for the 30 mM concentration case, the measured amplitude of the detected signal of the sensor does not show the previously described behavior. This curve shows a monotonous and smooth decrease with time, with the lowest initial slope or precipitation rate, and no plateau regime is observed for the same time window (500 s) used for the other two concentration cases. In fact, the signal decrease ratio is now −0.13 µV/s (300 s < time window < 500 s), just slightly lower than in the initial step of the deposition process.

Finally, the change in the detected magnetoelastic resonance frequency, and the subsequently determined mass load, for the three solution concentrations presented in this study are shown in Figure 5. The deposited mass values were calculated from the resonance frequencies by applying the calibration expression given in Equation (3), using the fitting coefficients from Figure 2. The error bars in the figure are calculated as:

$$\varepsilon_{\Delta m} = \sqrt{\left(\frac{\partial \Delta m}{\partial a_1}\varepsilon_{a_1}\right)^2 + \left(\frac{\partial \Delta m}{\partial a_2}\varepsilon_{a_2}\right)^2 + \left(\frac{\partial \Delta m}{\partial \Delta f}\varepsilon_f\right)^2} \tag{5}$$

where ε_{a_1} and ε_{a_2} are the errors obtained by the fitting procedure in Figure 2, and $\varepsilon_{\Delta f}$ is the experimental uncertainty in the resonance frequency ($\varepsilon_{\Delta f} = 100$ Hz).

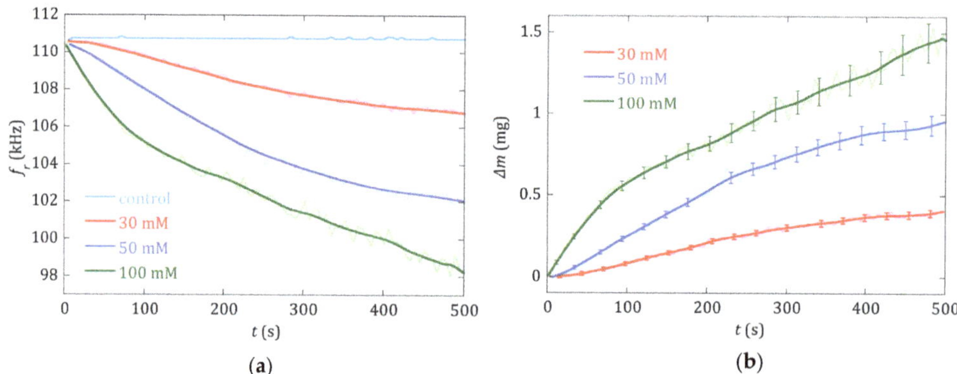

Figure 5. (**a**) Change of the magnetoelastic resonance frequency measured during the precipitation process for different reactant concentrations (30, 50 and 100 mM) and for the control test curve (sensor in a vial with distilled water). (**b**) Deposited mass of calcium oxalate crystals on the sensor during the reaction of precipitation for different reactant concentrations (30 mM, 50 mM and 100 mM), with the mass determination error. The actual measurements are plotted in light color. Intense solid curves correspond to smoothed data.

Figure 5 shows curves that evolve in a monotonous way in all cases, showing that the increase of the deposited mass of calcium oxalate onto the magnetoelastic strip leads to a decrease in its measured resonance frequency. However, the evolution of the curves appearing in Figures 4 and 5 is different, and hints at the complementarity of the information given by each one: while the decrease of the amplitude of the magnetoelastic resonance in the detected signal gives information about how fast this process occurs and eventually finishes, the decrease in the measured magnetoelastic resonance frequency value gives information about the amount of mass deposited along this process, by using the corresponding calibration equation.

Related to the practical application of our detection system, a targeted calcium oxalate concentration of 100 μmol/L = 12.8 mg/L (considering this value as a risk value for forming stones) translates to 0.015 mg = 15 μg of CaC_2O_4 to be detected in our 1.2 mL solution vial. According to the sensor calibration previously obtained (Section 2.2), a mass load of 0.015 mg of calcium oxalate on the resonator produces a magnetoelastic resonance frequency change of 0.15 kHz, just in the theoretical limit of the detectable change for our experimental set-up. Considering the minimum change in resonance frequency that we can discern (100 Hz), and the calibration constants obtained, the minimum theoretical deposited mass that our system is capable of detecting is 0.01 mg = 10 μg. However, the noise of the measured resonant frequency curves, up to 1 kHz (Figure 5a), raises this detection limit to 0.1 mg = 100 μg.

The results presented in this study demonstrate the feasibility of using magnetoelastic sensors in those cases where remote and non-destructive detection is required. This is the case for most biological or chemical agent detections. It turns out to be also a fast detection technique, that can provide accurate results in short time windows, of the order of several minutes. These rapid measurements are useful for monitoring this type of precipitation reaction, which can be quite fast, with sufficient accuracy.

Future work has to address the increase of the sensitivity of our magnetoelastic sensor, by reducing both the resonance frequency discrimination (below 100 Hz) and the noise in these measurements. Following this line of action, we will be able to quickly quantify the deposition of such inorganic salts with a resolution of, at least, below 10 μg.

4. Conclusions

The magnetoelastic resonance is an adequate phenomenon to be used in order to monitor, in real-time, the precipitation reaction of physiological inorganic salts, such as calcium oxalate (CaC_2O_4). The mass deposition of this salt onto the surface of the strip of the resonator gives rise to a shift of the resonance frequency that, by using the corresponding calibration curve, allows one to determine the mass amount deposited onto the resonator. Complementary information of the precipitation process arises from the measurement of the signal given by the amplitude of the detected magnetoelastic resonance, which is able to determine the time window in which the salt precipitation process occurs.

The corrosion resistant alloy used in this experiment ($Fe_{73}Cr_5Si_{10}B_{12}$) has been demonstrated to be as capable as the commercial Metglas® 2826 alloy in monitoring precipitation reactions, with the advantage that no pre-treatment is required to prevent oxidation when used as a precipitation reaction sensor. In addition, the aspect ratio has been shown to play an important role in the sensitivity of the magnetoelastic sensor, leading to a more mass-sensitive sensor that allows the detection of smaller amounts of precipitate, and therefore a more accurate understanding of this kind of processes.

Author Contributions: A.S.S. and J.G. conceived and designed the work and needed measurements; A.S.S. and B.S. performed the experiments; J.G. and A.G.-A. wrote the manuscript. All authors analyzed the data and discussed the results and implications, and commented on the manuscript at all stages. All authors read and approved the final manuscript.

Funding: The authors would like to thank the financial support from the Basque Government under μ4indust and IDEA projects (KK-2019/00101 and KK-2019/00039, Elkartek program) and University Basque Research Groups Funding (IT1245-19).

Acknowledgments: Technical and human support provided by SGIker (UPV/EHU, MICINN, GV/EJ, ESF) is gratefully acknowledged.

Conflicts of Interest: The authors declare no conflict of interest.

References

1. Luborsky, F.E. Amorphous ferromagnets. In *Ferromagnetic Materials*; Wohlfart, E.P., Ed.; Elsevier: Amsterdam, The Netherlands, 1980; Volume 1, ISBN 0-444-85311-1.
2. Squire, P.T. Magnetomechanical measurements of magnetically soft amorphous materials. *Meas. Sci. Technol.* **1994**, *5*, 67–81. [CrossRef]
3. Marín, P.; Marcos, M.; Hernando, A. High magnetomechanical coupling on magnetic microwire for sensors with biological applications. *Appl. Phys. Lett.* **2010**, *96*, 262512. [CrossRef]
4. Grimes, C.A.; Mungle, C.S.; Zeng, K.; Jain, M.K.; Dreschel, W.R.; Paulose, M.; Ong, G.K. Wireless magnetoelastic resonance sensors: A critical review. *Sensors* **2002**, *2*, 294–313. [CrossRef]
5. Stoyanov, P.G.; Grimes, C.A. A remote query magnetostrictive viscosity sensor. *Sens. Actuator A Phys.* **2000**, *80*, 8–14. [CrossRef]
6. Bouropoulos, N.; Kouzoudis, D.; Grimes, C. The real-time, in situ monitoring of calcium oxalate and brushite precipitation using magnetoelastic sensors. *Sens. Actuators B Chem.* **2005**, *109*, 227–232. [CrossRef]
7. Singh, A.K. Kidney Stones. In *Decision Making in Medicine*, 3rd ed.; Mushlin, S.B., Greene, H.L., Eds.; Elsevier: Amsterdam, The Netherlands, 2010; pp. 364–367. [CrossRef]

8. Curhan, G.C. Section XI: Renal and Genitourinary Diseases, Chapter 128 Nephrolithiasis. In *Goldman's Cecil Medicine*, 24th ed.; Elsevier: Philadelphia, PA, USA, 2012; Volume 1, pp. 789–794. ISBN 978-1-4377-1604-7.

9. Pak, C.Y.; Sakhaee, K.; Moe, O.W.; Poindexter, J.; Adams-Huet, B. Defining hypercalciuria in nephrolithiasis. *Kidney Int.* **2011**, *80*, 777–782. [CrossRef] [PubMed]

10. Robertson, W.G.; Peacock, M. The cause of idiopathic calcium stone disease: Hypercalciuria or hyperoxaluria? *Nephron* **1980**, *26*, 105–110. [CrossRef] [PubMed]

11. Massey, L.K.; Roman-Smith, H.; Sutton, R.A. Effect of dietary oxalate and calcium on urinary oxalate and risk of formation of calcium oxalate kidney stones. *J. Acad. Nutr. Diet.* **1993**, *93*, 901–906. [CrossRef]

12. Hallson, P.C.; Rose, A. Chemical measurement of calcium oxalate crystalluria: Results in various causes of calcium urolithiasis. *Urol. Int.* **1990**, *45*, 332–335. [CrossRef] [PubMed]

13. Robertson, W.G.; Scurr, D.S.; Bridge, C.M. Factors influencing the crystallisation of calcium oxalate in urine-critique. *J. Cryst. Growth* **1981**, *53*, 182–194. [CrossRef]

14. Stanković, A.; Šafranko, S.; Kontrec, J.; Njegić-Džakula, B.; Lyons, D.M.; Marković, B.; Kralj, D. Calcium Oxalate Precipitation in Model Systems Mimicking the Conditions of Hyperoxaluria. *Cryst. Res. Technol.* **2019**, *54*, 1800210. [CrossRef]

15. Sagasti, A.; Lopes, A.C.; Lasheras, A.; Palomares, V.; Carrizo, J.; Gutiérrez, J.; Barandiarán, J.M. Corrosion resistant metallic glasses for biosensing applications. *AIP Adv.* **2018**, *8*, 047702. [CrossRef]

16. Sagasti, A.; Palomares, V.; Porro, J.M.; Orúe, I.; Sánchez-Ilarduya, M.B.; Lopes, A.C.; Gutiérrez, J. Magnetic, Magnetoelastic and Corrosion Resistant Properties of (Fe–Ni)-Based Metallic Glasses for Structural Health Monitoring Applications. *Materials* **2020**, *13*, 57. [CrossRef] [PubMed]

17. Sagasti, A.; Gutiérrez, J.; Lasheras, A.; Barandiarán, J.M. Size dependence of the magnetoelastic properties of metallic glasses for actuation applications. *Sensors* **2019**, *19*, 4296. [CrossRef] [PubMed]

18. Gutiérrez, J.; Lasheras, A.; Martins, P.; Pereira, N.; Barandiaran, J.M.; Lanceros-Mendez, S. Metallic glass/PVDF magnetoelectric laminates for resonant sensors and actuators: A review. *Sensors* **2017**, *17*, 1251. [CrossRef] [PubMed]

19. Landau, L.D.; Lifshitz, E.M. Elastic waves. In *Theory of Elasticity*; Oxford Pergamon Press: Oxford, UK, 1975; p. 116.

20. Savage, H.; Abbundi, R. Perpendicular susceptibility, magnetomechanical coupling and shear modulus in $Tb_{0.27}Dy_{0.73}Fe_2$. *IEEE Trans. Mag.* **1978**, *14*, 545–547. [CrossRef]

21. Metglas®Inc. Magnetic Materials. Available online: https://metglas.com/magnetic-materials/ (accessed on 10 December 2019).

22. Metglas®Inc. Available online: https://metglas.com/wp-content/uploads/2016/12/Metglas-Alloy-2826MB3-Iron-Nickel-Based-Alloy.pdf (accessed on 27 January 2020).

23. Sagasti, A. Functionalized Magnetoelastic Resonant Platforms for Chemical and Biological Detection Purposes. Ph.D. Thesis, University of the Basque Country (UPV/EHU), Leioa, Spain, March 2018.

24. Sagasti, A.; Gutiérrez, J.; San Sebastián, M.; Barandiarán, J.M. Magnetoelastic Resonators for Highly Specific Chemical and Biological Detection: A Critical Study. *IEEE Trans. Mag.* **2017**, *53*, 4000604. [CrossRef]

25. Schmidt, S.; Grimes, C.A. Characterization of nano-dimensional thin-film elastic moduli using magnetoelastic sensors. *Sens. Actuator A Phys.* **2001**, *94*, 189–196. [CrossRef]

26. Aqion. Available online: https://www.aqion.de/site/168 (accessed on 16 March 2020).

27. Pankow, J.F. *Aquatic Chemistry Concepts*, 2nd ed.; CRC Press: Boca Raton, FL, USA, 2019; p. 220.

28. Xyla, A.G.; Koutsoukos, P.G. Effect of diphosphonates on the precipitation of calcium-carbonate in aqueous solutions. *J. Chem. Soc. Faraday Trans. 1 F* **1987**, *83*, 1477–1484. [CrossRef]

MDPI

St. Alban-Anlage 66

4052 Basel

Switzerland

Tel. +41 61 683 77 34

Fax +41 61 302 89 18

www.mdpi.com

Sensors Editorial Office

E-mail: sensors@mdpi.com

www.mdpi.com/journal/sensors